HUMAN–COMPUTER INTERACTION, 1997, Volume 12, pp. 301–309

Introduction to This Special Issue on Cognitive Architectures and Human–Computer Interaction

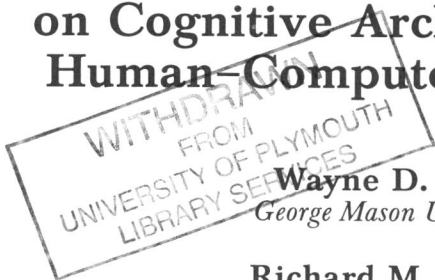

Wayne D. Gray
George Mason University

Richard M. Young
University of Hertfordshire

Susan S. Kirschenbaum
Naval Undersea Warfare Center Division, Newport, RI

his special issue was assembled by editors and contributors who believe
at cognitive architectures provide the most important new contribution to
theoretical basis for HCI (human–computer interaction) since the publi-
ion of *The Psychology of Human–Computer Interaction* (Card, Moran, &
vell, 1983). In this introduction, we provide a brief overview of what
itive architectures are and why we find them exciting. Then we intro-
the four architectures represented by articles in this special issue.

. **Gray** is a cognitive scientist with an interest in how artifact design
cognition required to perform tasks; he is Chair of the Human
nd Applied Cognition Program at George Mason University.
M. **Young** works in the Department of Psychology at the University
dshire; he is interested in cognitive modeling and its application to
h omputer interaction. **Susan S. Kirschenbaum** is an an engineering
psy ogist in the Department of Combat Systems at the Naval Undersea
Warfare Center Division, Newport, RI, where she studies and models expert
submarine decision-makers.

WHAT IS A COGNITIVE ARCHITECTURE?

According to Howes and Young (in this special issue), "A cognitive architecture embodies a scientific hypothesis about those aspects of human cognition that are relatively constant over time and relatively independent of task." A sufficiently complete cognitive architecture will ensure that theories of, say, reading and comprehension rely on the same mechanisms of information processing, storage, and retrieval that have been shown to be valid for playing chess, handling air traffic control, doing transcription typing, using a hand calculator, and programming a VCR. Rather than postulating a different set of mechanisms and parameters for each phenomenon studied, a cognitive architecture attempts to apply the same core set of constructs across an entire range of phenomena. Currently, all cognitive architectures remain research projects. All are incomplete.

Cognitive architectures differ from traditional cognitive theorizing in three important ways. First, traditional cognitive theories focus on building microtheories of isolated phenomena or mechanisms (Newell, 1973), whereas cognitive architectures seek integration. Second, attempts to show how theoretically distinct mechanisms combine, interleave, or contribute to complex cognition lead many designers of cognitive architecture to emphasize the control structure of cognition. Third, and perhaps most relevant for HCI, this emphasis on control structure has led most cognitive architectures to be cast in a form amenable to computer implementation, so that individual models built within the architecture take the form of computational cognitive models.

Constructing a model within one of these cognitive architectures involves writing a program in a powerful, restricted, and specialized programming language. The languages must be powerful if they are to enable the modeler (i.e., the programmer) to build models that can emulate even a fraction of the tasks performed by human cognition. The languages are restricted. Whereas traditional programming languages are designed to give the programmer great flexibility in implementing an algorithm, the point of a computational cognitive model is that the data structures and algorithms used are constrained by cognitive theory. We are not talking about artificial intelligence programming, in which the goal is, for example, to write a computer program that will beat Garry Kasparov at chess. This is computational cognitive modeling, in which the goal is to write a computer program that will play chess the same way Garry Kasparov does. The former has been done; the latter remains a significant challenge. Finally, the programming languages of cognitive architectures are specialized insofar as each was developed to emphasize certain aspects of human cognition. Although the goal may be a unified theory of all human cognition, there are many paths leading toward that goal, and, from different starting points, each current cognitive architecture follows a different route.

WHY ARE COGNITIVE ARCHITECTURES IMPORTANT FOR HCI?

Cognitive architectures are in the process of becoming the preferred route for bringing cognitive theory to bear on HCI. Several of their strengths provide help for meeting the challenge of understanding, evaluating, and engineering the interactions between humans and their information artifacts:

- Cognitive architectures provide the mechanisms and processes by which different aspects of cognition—attention, memory, problem solving, learning, and so forth—work in concert to exhibit integrated behavior of the kind required to perform tasks with interactive computer interfaces.
- This degree of integration provides a starting point for analyzing (and then resynthesizing) the cognition that occurs at the interface. When confronted with an interface, the cognitive modeler asks, "How would C–I [or EPIC, Soar, ACT–R, etc.] perform the task using this design?" The answer is cast in terms of a complete theory (a working model) of all the cognition required to do the task using the interface in question. In contrast, traditional methods of cognitive theorizing tend to be phenomenon based (Newell, 1973). Narrow-scope theories of divided attention or decision making—or whatever—focus on those elements of the interaction that best illustrate their defining phenomena, whereas other elements tend to be ignored.
- Computational models derived from cognitive architectures can be applied to tasks involving both routine skill and problem solving. Skilled cognition is contrasted to that requiring problem solving, with the distinction involving the amount of control knowledge used to guide search through a problem space (Card et al., 1983). Whether a particular task is routine will of course depend on the knowledge and experience of the individual, but typical examples of skilled tasks include transcription typing (for a skilled typist) and copying files (for a skilled computer user). Examples of tasks involving a large component of problem solving include computer programming, scientific reasoning, and most design activities. Many tasks involve components of both skilled cognition and problem solving. Unlike most other approaches to cognitive modeling, cognitive architectures can handle tasks across the continuum.
- Computational cognitive models based on cognitive architectures are generative in that the same model can be applied to a range of different tasks or variations on a task. This contrasts with more descriptive analyses, for which new tasks, or variations in performing an already modeled task, require the development of new descriptive models.

- Finally, there is the issue of learning. Although static computational models can be used to measure what must be learned and to predict learning time (e.g., Bovair, Kieras, & Polson, 1990), only models based on cognitive architectures can model the learning process itself. Hence, architecture-derived computational models can be used to determine how different tasks or instructions lead to different types and amounts of learning and how forgetting or other causes of memory failure underlie the difficulties encountered by casual users of complex systems.

Besides providing a conduit for applying theory to design, cognitive architectures are also on the path driving the development of cognitive theory from applied problems in HCI. Interacting with modern computer systems requires the user to respond to text and graphic events as they occur. It also requires a complex orchestration of cognition with perception and motor processes. This challenge to cognitive theory to account for the event-driven, display-based performance required by graphical user interfaces (GUIs) and multimedia interactions has brought renewed attention to HCI phenomena from the cognitive science community. These researchers are looking at our field with new eyes and are finding that minor differences in how an interface is designed can lead to significant differences in the interplay among cognition, perception, and motor activities and thence to large differences in performance. For example, studies of decision making have shown that the cognitive cost of processing (Kleinmuntz & Schkade, 1993; Payne, Bettman, & Johnson, 1993) or accessing information (e.g., by eye movements vs. moving a mouse to its location; Lohse & Johnson, 1996) leads the user to adopt different decision-making strategies that vary in their effectiveness.

In the short run, this influx of basic researchers into HCI means that most articles written about the application of cognitive architectures will be of more interest to researchers than to practitioners. In the long run, however, practitioners will benefit. Driven by the needs of the research community, HCI will become the first applied field in which phenomena can be fully accounted for by cognitive architectures. Furthermore, the skills possessed by the HCI community are exactly those required to take such research tools and transform them into tools for practitioners. The day of the routine use of cognitive architectures to model HCI tasks may not yet be upon us, but it is not far away.

THE ARCHITECTURES AND ARTICLES IN THIS SPECIAL ISSUE[1]

Until about the mid-1980s, it was possible to identify only one candidate for a cognitive architecture—ACT* (Anderson, 1983). Today, we can list about half a dozen. Of the cognitive architectures that have emerged since 1983, Soar (Newell, 1990) is the most established and best known within the HCI community. Unlike the ACT family, Soar has always had a programming language as an integral part of the theory. This has meant that, since its inception, to theorize in Soar has entailed writing a computational cognitive model to perform the behavior in question. One criterion for a successful theory is its acceptance by researchers outside its circle of developers. On this basis, Soar clearly leads the pack, as well-established communities of Soar modelers exist throughout the United States and Europe.

In this special issue, Howes and Young use Soar as a springboard to explore how an architecture constrains the way that models of behavior must be constructed. Howes and Young illustrate their points with the seemingly roundabout way that Soar models learn to associate menu labels with actions—that is, how Soar learns to use a GUI. An interesting and important conclusion is that, although the natural way to implement methods for this task in Soar seem very indirect from a traditional programming perspective, the models appear to capture the essence of human performance.

The construction–integration (C–I) architecture started life as a theory of language comprehension (Kintsch, 1988). Developments by Mannes and Kintsch (1991) as well as Doane, Pellegrino, and Klatsky (1990) transformed C–I into a tool for modeling HCI tasks. In this special issue, Kitajima and Polson present the latest installment of their LICAI (linked model of comprehension-based action planning and instruction taking) extensions to C–I theory. As with previous versions (Kitajima, 1989; Kitajima & Polson, 1992, 1995, 1996), the current article focuses on the cognition involved when an experienced GUI user attempts to learn a new software package. An interesting contribution of the article is its development and elaboration of mechanisms for goal selection within a C–I architecture.

EPIC (executive process–interactive control) was developed by Meyer and Kieras (1997a, 1997b) for the stated purpose of representing embodied cognition. Compared to other cognitive architectures, EPIC is an idiot

---

1. Further cognitive architectures that are relevant to HCI but that are not represented in this special issue include CAPS (e.g., Byrne & Bovair, 1997), ZIPPY (e.g., Rist, 1995), and ICS (e.g., Barnard & May, 1993; May, Barnard, & Blandford, 1993).

savant: Its cognitive component is minimized in favor of well-developed motor and perceptual components. This intentionally lopsided development permits EPIC modelers to determine constraints on cognitive performance that emerge from properties of the human's input (auditory and visual perception) and output (speech and motor movement) mechanisms. A modeler using EPIC must adhere to the parameters provided for moving eyes, processing vision, and so forth. In this special issue, Kieras and Meyer present HCI researchers with a detailed introduction to EPIC. In addition, they present data and model predictions for several tasks of interest to HCI researchers, including a menu search task originally studied by Nilsen (1991; see also Hornof & Kieras, 1997).

The original ACT* architecture (Anderson, 1983) metamorphized first into ACT–R (Anderson, 1993) and now into ACT–R 4.0 (Anderson, in press). Along the way, it has expanded in coverage, acquired a fully implemented programming language, and grown eyes and fingers. Indeed, in this special issue, Anderson, Matessa, and Lebiere argue that the newly augmented ACT–R is now fully capable of handling the cognition that occurs at the computer interface. Their claim is that ACT–R now has a theory of visual perception (discussed in detail elsewhere) and visual attention (the focus of their current article). These two components expand ACT–R's coverage so that it can now be used to model human interaction with computer applications. The validity of the claim is demonstrated by the application of ACT–R to several classic but never before modeled psychological paradigms.

One of the tasks modeled by Anderson, Matessa, and Lebiere is the menu search task (Nilsen, 1991) modeled by Kieras and Meyer. After developing an ACT–R model, Anderson et al. apply it to a new menu search task. They claim that ACT–R predicts human performance in this new task using the same parameters and assumptions required to model the first task. They further claim that the model developed by Kieras and Meyer would need additional assumptions to fit the new data. This comparison provides an interesting example of two different architectures tackling the same benchmark task. Such competitive argumentation (VanLehn, Brown, & Greeno, 1982) or cooperative analysis across shared scenarios (Young & Barnard, 1987) may offer the best ways to compare the relative merits of different cognitive architectures.

CAVEATS AND CONCLUSIONS

In this introduction, we have argued for the importance of cognitive architectures as a vehicle for injecting cognitive theory into HCI. However, the critical importance of cognitive architectures is far from universally accepted, and we do not expect to convince those who believe that cognitive theory has no part to play in HCI. Moreover, despite our espousal of cognitive architectures for HCI, we must acknowledge two important caveats.

First, by focusing on cognitive theory, we do not mean to imply that other behavioral and social sciences have no role in HCI. For example, the research community on computer-supported cooperative work (CSCW) brings a variety of cognitive and noncognitive sciences to bear on HCI issues. Some of these issues are outside the scope of cognitive theories. In addition, some of the cognitive issues addressed are closely intertwined with social issues in ways that are not currently amenable to architecture-based approaches. Second, there are basic cognitive processes that are important to HCI but that are not sufficiently well understood to be incorporated into existing cognitive architectures. Clear examples of such areas come from speech understanding and visual perception.

Although these caveats temper our claims, they do not temper our enthusiasm. Despite a gestation period that began in the 1950s (Simon, 1991, chap. 12), the advent of runnable cognitive architectures is a relatively recent occurrence. For example, between 1983 and 1993, 11 articles either foreshadowed or applied something like an architecture-based approach to HCI phenomena. In contrast, since 1993, 21 articles (including the 4 in this special issue) have built architecturally inspired models of HCI tasks. We see cognitive architectures as a new phenomenon—one that is gaining wide acceptance within the cognitive science community and whose potential for application to HCI concerns is just beginning to be appreciated.

NOTES

Background. The articles in this special issue have their origins in a workshop conducted at the CHI'95 Conference on Human Factors in Computing Systems, Denver, CO (Kirschenbaum, Gray, & Young, 1996).

Support. Wayne D. Gray's work on this special issue was supported in part by Office of Naval Research Grant N00014–95–1–0175 and a Fellowship from the Krasnow Institute for Advanced Studies. Susan S. Kirschenbaum's work on this special issue was supported in part by the Office of Naval Research (Project A10328) and the Naval Undersea Warfare Center's Independent Research Program (Project E10207).

Editors' Present Addresses. Wayne D. Gray, George Mason University, MSN 3f5, Fairfax, VA 22030. E-mail: gray@gmu.edu. Richard M. Young, Department of Psychology, University of Hertfordshire, Hatfield, Herts, AL10 9AB, United Kingdom. E-mail: r.m.young@herts.ac.uk. Susan S. Kirschenbaum, Code 2214, Building 1171/1, Naval Undersea Warfare Center Division, Newport, RI 02841. E-mail: kirsch@c223.npt.nuwc.navy.mil.

REFERENCES

Anderson, J. R. (1983). *The architecture of cognition.* Cambridge, MA: Harvard University Press.

Anderson, J. R. (1993). *Rules of the mind.* Hillsdale, NJ: Lawrence Erlbaum Associates, Inc.

Anderson, J. R. (in press). *Atomic components of thought.* Mahwah, NJ: Lawrence Erlbaum Associates, Inc.

Barnard, P. J., & May, J. (1993). Cognitive modeling for user requirements. In P. Byerley, P. J. Barnard, & J. May (Eds.), *Computers, communication and usability: Design issues, research and methods for integrated services* (pp. 101–146). Amsterdam: North-Holland.

Bovair, S., Kieras, D. E., & Polson, P. G. (1990). The acquisition and performance of text-editing skill: A cognitive complexity analysis. *Human–Computer Interaction, 5,* 1–48.

Byrne, M. D., & Bovair, S. (1997). A working memory model of a common procedural error. *Cognitive Science, 21,* 31–61.

Card, S. K., Moran, T. P., & Newell, A. (1983). *The psychology of human–computer interaction.* Hillsdale, NJ: Lawrence Erlbaum Associates, Inc.

Doane, S. M., Pellegrino, J. W., & Klatsky, R. L. (1990). Expertise in a computer operating system: Conceptualization and performance. *Human–Computer Interaction, 5,* 267–304.

Hornof, A. J., & Kieras, D. E. (1997). Cognitive modeling reveals menu search is both random and systematic. *Proceedings of the CHI'97 Conference on Human Factors in Computing Systems,* 107–114. New York: ACM.

Kintsch, W. (1988). The role of knowledge in discourse comprehension: A construction–integration model. *Psychological Review, 95,* 163–182.

Kirschenbaum, S. S., Gray, W. D., & Young, R. M. (1996). Cognitive architectures and HCI [Workshop report]. *SIGCHI Bulletin, 28*(2), 18–21.

Kitajima, M. (1989). A formal representation system for the human–computer interaction process. *International Journal of Man–Machine Studies, 30,* 669–696.

Kitajima, M., & Polson, P. G. (1992). A computational model of skilled use of a graphical user interface. *Proceedings of the CHI'92 Conference on Human Factors in Computing Systems,* 241–249. New York: ACM.

Kitajima, M., & Polson, P. G. (1995). A comprehension-based model of correct performance and errors in skilled, display-based, human–computer interaction. *International Journal of Human–Computer Studies, 43,* 65–99.

Kitajima, M., & Polson, P. G. (1996). A comprehension-based model of exploration. *Proceedings of the CHI'96 Conference on Human Factors in Computing Systems,* 324–331. New York: ACM.

Kleinmuntz, D. N., & Schkade, D. A. (1993). Information displays and decision processes. *Psychological Science, 4,* 221–227.

Lohse, G. L., & Johnson, E. J. (1996). A comparison of two process tracing methods for choice tasks. *Organizational Behavior and Human Decision Processes, 68*(1), 28–43.

Mannes, S., & Kintsch, W. (1991). Routine computing task: Planning as understanding. *Cognitive Science, 15,* 305–342.

May, J., Barnard, P. J., & Blandford, A. E. (1993). Using structural descriptions of interfaces to automate the modeling of user cognition. *User Modeling and User-Adapted Interaction, 3,* 27–64.

Meyer, D. E., & Kieras, D. E. (1997a). A computational theory of executive cognitive processes and multiple-task performance: Part 1. Basic mechanisms. *Psychological Review, 104,* 3–65.

Meyer, D. E., & Kieras, D. E. (1997b). A computational theory of executive cognitive processes and multiple-task performance: Part 2. Accounts of psychological refractory-period phenomena. *Psychological Review, 104,* 749–791.

Newell, A. (1973). You can't play 20 questions with nature and win: Projective comments on the papers of this symposium. In W. G. Chase (Ed.), *Visual information processing* (pp. 283–308). New York: Academic.

Newell, A. (1990). *Unified theories of cognition.* Cambridge, MA: Harvard University Press.

Nilsen, E. L. (1991). *Perceptual-motor control in human–computer interaction* (Technical Report 37). Ann Arbor: University of Michigan, Cognitive Science and Machine Intelligence Laboratory.

Payne, J. W., Bettman, J. R., & Johnson, E. J. (1993). *The adaptive decision maker.* New York: Cambridge University Press.

Rist, R. S. (1995). Program structure and design. *Cognitive Science, 19,* 507–562.

Simon, H. A. (1991). *Models of my life.* New York: Basic.

VanLehn, K., Brown, J. S., & Greeno, J. (1982). Competitive argumentation in computational theories of cognition. In W. Kintsch, J. Miller, & P. Polson (Eds.), *Methods and tactics in cognitive science.* Hillsdale, NJ: Lawrence Erlbaum Associates, Inc.

Young, R. M., & Barnard, P. (1987). The use of scenarios in human–computer interaction research: Turbocharging the tortoise of cumulative science. *Proceedings of the CHI+GI'87 Conference on Human Factors in Computing Systems and Graphics Interface,* 291–296. New York: ACM.

ARTICLES IN THIS SPECIAL ISSUE

Anderson, J. R., Matessa, M., & Lebiere, C. (1997). ACT–R: A theory of higher level cognition and its relation to visual attention. *Human–Computer Interaction, 12,* 311–343.

Howes, A., & Young, R. M. (1997). The role of cognitive architecture in modeling the user: Soar's learning mechanism. *Human–Computer Interaction, 12,* 345–389.

Kieras, D. E., & Meyer, D. E. (1997). An overview of the EPIC architecture for cognition and performance with application to human–computer interaction. *Human–Computer Interaction, 12,* 391–438.

Kitajima, M., & Polson, P. G. (1997). A comprehension-based model of exploration. *Human–Computer Interaction, 12,* 439–462.

HUMAN–COMPUTER INTERACTION, 1997, Volume 12, pp. 311–343

The Role of Cognitive Architecture in Modeling the User: Soar's Learning Mechanism

Andrew Howes
Cardiff University of Wales

Richard M. Young
University of Hertfordshire

ABSTRACT

What is the role of a cognitive architecture in shaping a model built within it? Compared with a model written in a programming language, the cognitive architecture offers theoretical constraints. These constraints can be "soft," in that some ways of constructing a model are facilitated and others made more difficult, or they can be "hard," in that certain aspects of a model are enforced and others ruled out. We illustrate a variety of these possibilities. In the case of Soar, its learning mechanism is sufficiently constraining that it imposes hard constraints on models constructed within it. We describe how one of these hard constraints deriving from Soar's learning mechanism ensures that models constructed within Soar must learn a display-based skill and, other things being equal, must find display-based devices easier to learn than keyboard-based devices. We discuss the relation between architecture and model in terms of the degree to which a model is "compliant" with the constraints set by the architecture. Although doubts are sometimes expressed as to whether

Andrew Howes is a Lecturer in the School of Psychology at the Cardiff University of Wales; he is interested in cognitive modeling and human–computer interaction. **Richard M. Young** is a Principal Lecturer in the Psychology Department at the University of Hertfordshire, Hatfield; he is interested in cognitive modeling and its application to human–computer interaction.

CONTENTS

cognitive architectures have any empirical consequences for user modeling, our analysis shows that they do. Architectures play their part by imposing theoretical constraints on the models constructed within them, and the extent to which the influence of the architecture shows through in the model's behavior depends on the compliancy of the model.

1. INTRODUCTION: COGNITIVE ARCHITECTURES AND USER MODELING

A cognitive architecture embodies a scientific hypothesis about those aspects of human cognition that are relatively constant over time and relatively independent of task. Examples range from architectures claiming broad scope, such as Soar (Newell, 1990) and ACT–R (Anderson, 1993), through ICS (interacting cognitive subsystems; Barnard, 1987), to more specialized architectures such as C–I (construction–integration; Kintsch, 1988). The different architectures in principle offer alternative, competing theories of cognition, but the considerable overlap in their structures and assumptions—and the fact that they tend to exhibit different areas of strength and weakness—makes any evaluative comparison between them a complex and difficult undertaking.

As reflected in this special issue of *Human–Computer Interaction* and in the workshop from which it stems (Kirschenbaum, Gray, & Young, 1996), a recent trend in work extending the state of the art in constructing psychological models of the computer user has been to cast the models within some chosen cognitive architecture (e.g., Anderson, Matessa, & Lebiere, 1997; Barnard & May, 1993; Howes & Young, 1996; Kitajima & Polson, 1995). In such models, part of the content of the model is supplied by the cognitive architecture itself; the rest is supplied by what the analyst must add to the (generic) architecture in order to construct a (specific) model. Later, we define that additional information as constituting the *model increment*. Thus, the theoretical content of the model is distributed between the cognitive architecture and the model increment. The contrast to working with an architecture is for the analyst to implement the user model directly in some programming language, such as Lisp or C, chosen for its ease of programming and its convenience for expressing the model but not for its theoretical contribution.

Building user models within a cognitive architecture in this fashion raises novel problems of practical and scientific evaluation. Given a model and a pattern of empirical evaluation against data and applications, how much of the credit for the successes, and blame for the failures, should be allocated to the architecture itself and how much to the model increment? More generally, the approach raises the question of how we should conceptualize, discuss, and assess the contribution of the cognitive architecture to the properties of the specific model.

In this article, we offer an approach to answering the question by exploring the notion of theoretical constraint. The core idea is that the cognitive architecture influences or biases the kinds of models that can be constructed within it by placing constraints on what can be done by the possible model increments. These constraints can be "soft," in that the construction of some kinds of model is facilitated by aspects of the architecture, whereas other kinds are made more difficult. Or the constraints can be "hard," in that certain features of the resulting models are enforced, whereas others are ruled out. To determine how much of the quality of a specific model and of its explanatory force should be attributed to the architecture and how much to the model increment, we need to determine the situations in which the architecture imposes hard constraints on models built within it.

In Section 2, as a necessary preliminary to considering cases of hard and soft constraints, we illustrate various ideas about model increments and the contents of models. Then, in Section 3, we present a case study, in which we examine the degree to which, and the ways in which, a particular feature (the learning mechanism) of a particular architecture (Soar) places hard constraints on all models that make use of the feature. In Section 4, we discuss the general lessons suggested by this study with regard to the use of cognitive architectures in user modeling. In Section 5, we draw conclusions.

2. THE ROLE OF THE ARCHITECTURE

To gain a better understanding of the issues raised by architecture-based models, we first (in Section 2.1) propose a framework for the information required to specify a particular model within a given cognitive architecture. We then (in Section 2.2) apply the framework to examine more closely the role of the model increment within a single architecture.

2.1. Specifying a Model Within a Cognitive Architecture

All models constructed within a given cognitive architecture share the features of that architecture. What distinguishes one model from another is the extra information that must be provided in order to define a particular model within the architecture. As we have already seen, we refer to this further specification as the *model increment*. (The rationale for the term becomes clearer later.) In this section, we look at the nature of the model increment in the case of four illustrative cognitive architectures. The purpose of this section is not to compare the architectures or make comparative judgements about their value but to present a framework for understanding architecture-based models and to demonstrate that the framework applies across a range of different architectures.

The form and content of the model increment varies from one architecture to another. In Soar (Newell, 1990), the model increment takes the form of a set of production rules that encode the relevant knowledge the user is postulated to have. The content of the rules can be about the task to be performed, the device and its interface, relevant background knowledge (e.g., the meaning of words; the order of days of the week), control information for methods of performing tasks, and indeed anything that may pertain to the user's behavior. The view adopted is that the Soar architecture, together with the knowledge expressed as the model increment, operating in a (usually simulated) task environment, yields a prediction of the user's behavior. Summarized schematically, we have

$$\text{Soar} + \text{Knowledge (of States, of Operators, ...)} + \text{Task Environment} \rightarrow \text{Behavior}$$

where the model increment consists of just the knowledge component.

The model increment for ACT–R (Anderson, 1993) is in many ways similar to that for Soar, although there is more diversity in the way information is expressed. First, knowledge is encoded in two different memories—a procedural memory, expressed as production rules, and a declarative memory, expressed as a network of linked nodes. Usually, simple factual knowledge is encoded in the declarative memory, and enactive knowledge about methods and procedures is encoded in the procedural memory. However, the analyst has a degree of discretion, so that declarative information about methods and procedures can be held in the declarative memory, and factual knowledge can be encoded in production rules that reflect the usage of that knowledge.[1] Second, a range of numerical parameters is open to specification in order to determine link strengths, initial activation levels, time constants, prior probabilities and estimates of costs and rewards, and so forth. Schematically, we have:

ACT–R + Declarative Knowledge + Procedural Knowledge
+ Numeric Parameters + Task Environment → Behavior

where the model increment consists of the two kinds of knowledge together with the parameter values.

The C–I architecture derives from a theory of sentence processing (Kintsch, 1988), so, not surprisingly, the information it requires to specify a model is primarily propositional in nature. Kitajima and Polson (1995) described the relevant knowledge under six headings—*task-goal, device-goal, display, long-term memory, candidate objects,* and *object–action pairs.* In addition, there are six numerical parameters for the model, describing aspects such as the relative weight to be given to the link between two propositions that share a common term, the multiplier for making links between the goals and the rest of the network stronger than other links, and the number of memory samples to be taken. Schematically, this yields:

C–I + Six Kinds of Propositions + Numeric Parameters
→ Action Choice

We omit the task environment from this formula, because C–I models are used to simulate extended sequences of interaction between user and device in a step-by-step fashion. A C–I model is run to examine a single action decision at a particular point in an interaction sequence. To get the prediction for the next decision, the goals and display part of the model increment are reset to describe the updated situation, and the model is run again to make a new prediction.

---

1. Anderson (1993) offered various arguments in favor of more varied architectural representational structures.

The ICS model (Barnard, 1987) presents a still different picture. Unlike the other architectures considered here, ICS is not a simulation architecture. Instead, it provides a structure and a set of concepts concerning the user's cognition, in terms of which the analyst can describe an interactive situation and argue to a prediction of its consequences for the user. In certain cases, the role of the analyst can be partly taken over by an expert system (Barnard & May, 1993; Barnard, Wilson, & MacLean, 1988), which receives help in mapping the situation into ICS terminology and then derives the consequences and reports the predictions. In these cases, the expert system is playing (part of) the role of analyst, not simulating the user. In order to apply ICS to a particular question, information must be specified about the dynamic configuration of the internal components of ICS; the degree of overlearning ("proceduralization") of aspects of relevant skills in a task and perceptual setting; the content of the memories; and the representation of perceptual encodings (May, Barnard, & Blandford, 1993). In terms of our schematic formula, we have:

ICS + Process Configuration + Proceduralization +
Memory Contents + Perceptual Representations +
Dynamic Control $\rightarrow$ Prediction of Attributes of User Behavior

Again, there is no separate task environment, as an ICS model is not a simulation, and information about the task environment is distributed among other parts of the specification.

These four examples do not exhaust the possibilities for the forms the model increment can take. Their commonalities, however, serve to illustrate that, in one way or another, and although the different influences can be distributed among the components in different ways, the resulting behavior of a model built within a cognitive architecture depends on:

- The architecture itself.
- Knowledge and other forms of specification added to identify a particular model.
- The task and the environment within which the model operates.

For any one of these aspects of the model (e.g., the architecture) to be able to affect the behavior, it must in some way constrain the influence of the other aspects (e.g., the model increment and the task environment). (If an aspect exerts no constraint, it is difficult to see how it can be playing a role in determining the behavior.) In Section 2.2, we consider the case of a particular cognitive architecture, Soar, and examine how it constrains the properties of individual Soar models.

2.2. Role of the Architecture

The decomposition of an architecture-based model into architecture + model increment + task environment raises problems of credit assignment. For a particular feature of the resulting behavior, how do we determine the component to which it is due? Such problems are especially relevant in the context of evaluating the model. If there is a discrepancy between the behavior of the model and empirical data, where do we place the "blame"? Does it mean that the architecture is wrong, that the model increment is wrong, or that there is some error in the way we have simulated the task environment? Correspondingly, when the agreement between model and data is good, how do we allocate the "credit"? Should the agreement strengthen our confidence in the architecture, or does it tell us only that the added knowledge, so far as it goes, seems right?

In this section, we explore further the way in which the cognitive architecture and a model increment work together to determine a model's behavior. The key idea introduced here is that the architecture itself can propose actions to be taken, which the model increment can concur with, or can modify, or can simply override. To prepare the groundwork for this discussion, we begin by reviewing in some detail Laird's (1986) work on "universal weak method" (UWM).

Universal Weak Method (UWM)

One of the (many) ways in which a cognitive architecture differs from being just a programming environment is that a cognitive architecture may be able to proceed with an "incomplete" model increment in a way that has no counterpart for ordinary programs. In other words, a programming environment or language (e.g., C) has no "ideas of its own" about what to do. It does nothing at all until it is told what to do by being given a program. That program must be "complete," at least in the parts that get executed; otherwise, if the execution reaches a part of the program that is undefined, it will stop with an error message. That need not be so for a cognitive architecture. Soar, in particular, will "run" even with a missing or highly incomplete model increment by relying heavily on the notion of default behavior. Given the description of a problem or task, Soar will set about trying to perform it, even without any information about how to do so. Of course, its behavior under these conditions is not very "intelligent," but it is far from trivial. For internal problem-solving, Soar will by default engage in a form of depth-first search.

Laird (1986) extended the basic Soar architecture with a set of default production rules to yield an architecture exhibiting what he termed *universal weak method*. In this context, a weak method (Newell, 1969) is a problem-solving method that is very general and requires little knowledge about the task. Examples are (a) means–ends analysis and (b) the generate-and-test process. Because a weak method demands little domain-depend-ent knowledge, it can be used in many domains, especially when "stronger," more domain-specific knowledge is unavailable. Laird showed that, by adding to UWM what he called *method increments* (small collections of production rules that encode information about the domain and task), he could generate most of the known weak methods.

To understand Laird's (1986) demonstration, we need first to consider Soar's normal behavior. When Soar is performing a task for which the model increment directly supplies the necessary knowledge, it works in a problem space in which, from its current state, it repeatedly chooses an operator and then applies that operator to the current state to get another state that it treats as its new current state. When there is uncertainty about which operator to apply, Soar drops into a subgoal, the purpose of which is to resolve the uncertainty by selecting one of the operators. Without any further information (i.e. with a null knowledge increment), UWM's default behavior is to resolve uncertainty between operators by evaluating each of the candidates, which it does by applying the operator to the current state in a separate context and seeing where it leads. This process generates, as we have noted, a form of depth-first search, which is itself one of the classical weak methods.

Consider now what happens if we add to the UWM a method incre-ment consisting of (a) task-domain information sufficient to evaluate states (e.g., on a simple numerical scale) and (b) an item of general control information telling Soar that an operator leading to a state with a higher numerical evaluation than another is preferable to the other. The effect of that method increment is that, when Soar is uncertain about the choice of operators and is considering the options, it will select as best the operator that leads from the current state to the adjacent state with the highest evaluation. What emerges is behavior described by the classical weak method of "steepest ascent hill-climbing." It should be noted that the basic "engine" driving behavior remains that of Soar/UWM, whereas what makes the behavior take the form of hill climbing is the method increment.

Of interest for our present purpose is Laird's (1986) observation that not all of the weak methods can be generated in this fashion. There exists a group of weak methods, clustered around breadth-first search and includ-ing best-first search, that do not lend themselves to description as incre-mental variations on UWM. The reason is not difficult to see. UWM's "own" behavior (i.e., depth-first search) is not only economical of memory demands, but the storage it does require exhibits a strong property of "locality" in both time and space. The information that must be stored in conjunction with a given state is generated at about the time that the state is the current state, concerns some other state adjacent in the search space, and can be discarded at about the same time that the state itself is abandoned. None of this is true for the "difficult" weak methods. Their "breadthness" means that they have heavy storage demands, scattered across the search space, that must persist over time. Consequently, they require a processing regime fundamentally at variance with that of the UWM, in which all existing states must be searched in order to find the next one to treat as current.

Now, Soar is a computationally universal architecture, so it is not actually impossible to get it to perform breadth-first search. To do this, however, the analyst must write a model increment that consists essentially of a self-contained program for breadth-first search or whatever. Unlike the case with hill climbing, in which the behavior emerges jointly from the architecture and the method increment with a distinct contribution being made by each, for breadth-first search the behavior is determined almost entirely by the model increment (i.e., the production rules given to Soar) with almost no contribution from the architecture. The model cannot be decomposed in a meaningful or useful way into UWM together with a method increment. The breadth-first program achieves its ends by brute force, as it were—neither taking advantage of nor making concessions to the properties of the architecture. It might be written almost as well in a standard programming language.

Before we leave this topic, we should note that, although the constraints of UWM are, in the terminology introduced earlier, soft rather than hard (in the sense that they do not make it literally impossible for Soar/UWM to exhibit, e.g., breadth-first search), they nevertheless clearly delineate, in some informal but important sense, model increments that cooperate with the architecture from those that fight it. A soft constraint is not necessarily a weak constraint. The contrast between the two classes of methods is clearly marked.

Compliant Models

What we take from this discussion of Laird's (1986) analysis is a subtly different conception of the relation between cognitive architecture and model increment than the one with which we began. The relation of model increment to architecture is not at all like that of program to programming language. It is not the case (at least for Soar) that the architecture simply provides a framework within which the model is "programmed." Instead, the architectural mechanisms can propose a course of action. The model increment can go along with that proposed action, add to the action (which is what is done by the method increment for most of the weak methods), oppose and override the architecture's proposals (which is what must happen to yield breadth-first search), or do anything in between. In this way, we can see the role of the model increment as being not so much to generate behavior, because the architecture will behave anyway, but rather to modify or modulate the behavior that would otherwise occur.

Models built within a cognitive architecture clearly vary in the extent to which they follow the direction set by the architecture. We refer to this property of the model as its *compliancy*. A compliant model (e.g., the model for hill climbing) allows the architecture to play a significant role in determining behavior. A noncompliant model (e.g., the model for breadth-first search) consistently overrules the architecture. (Young, 1982, also defined *compliancy,* but our definition is considerably more precise; note that the idea goes back at least to the "natural methods" notion introduced by Moore & Newell, 1974.)

For Soar, at least, this idea of compliancy can be translated into concrete terms. Because Soar, no matter how little or how much knowledge it is given about a task, is usually in a position to behave, it always has proposals for the selection and sequencing of actions—what steps should be taken and in what order. A model is compliant to the extent that it follows those proposals. In the extreme case, Soar offers its "default" behavior, as already described. If the model increment defines a set of task-relevant possible moves but does not provide control information (i.e., adds no knowledge about the selection and sequencing of the moves), then the default behavior will be exhibited. A highly compliant model increment will, like the method increment for hill climbing, add just a little control information that occasionally directs the processing in a direction determined by the model, but most of the time the decisions about selection and sequencing can still be left to the architecture. Models that are less compliant will themselves specify more and more about the processing, leaving less and less to the architecture, whereas a highly noncompliant model provides essentially a complete algorithm for the desired behavior, permitting no influence to the architecture.

3. SOAR'S LEARNING MECHANISM

In this section, we continue our investigation of the relation between cognitive architecture and model increment by examining cases in which the architecture imposes hard constraints on the possible models. We consider several empirical regularities describing how people interact with computers and several models constructed in the Soar cognitive architecture, and we attempt to determine whether the behavior of the model is a consequence of the architecture or of the encoded knowledge (i.e., of the particular model increments). In particular, we examine the role of the aspect of Soar's learning mechanism called *backtracing.* Due to this mechanism, Soar models necessarily exhibit certain behaviors seen in HCI but must be configured with particular model increments in order to capture

Figure 1. Chunking encodes an episodic memory of problem-solving activity. Left panel: before chunking; Right panel: after chunking. *Note.* From "Learning consistent, interactive and meaningful device methods: A computational model," by A. Howes & R. M. Young, 1996, *Cognitive Science, 20,* p. 315. Copyright © 1996 by Cognitive Science Society, Inc. Reprinted with permission.

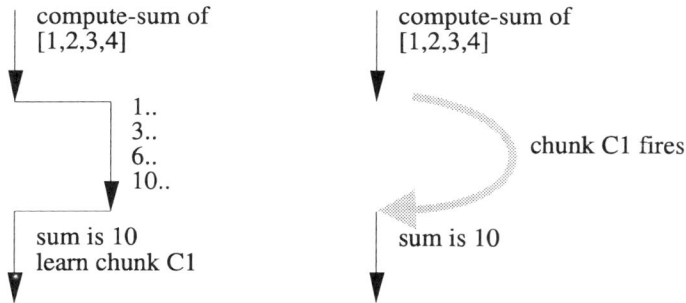

others. We draw on several examples taken from a model of the acquisition of task-action mappings (Howes & Young, 1996).

Soar's learning mechanism is known as *chunking*—an explanation-based mechanism in Mitchell's (1986; see Rosenbloom & Laird, 1986) sense. This mechanism learns by finding the conditions for why a particular conclusion was drawn and then creating new rules, or chunks, that propose the conclusion given the conditions.

Chunking can be viewed as a mechanism that caches the result of problem solving so that, on future occasions, the result can be retrieved without further processing. Consider the example of the task of computing the sum of a list of numbers—"Compute-sum of [1,2,3,4]." Imagine that the algorithm used consists of starting with a sum of $S = 0$ and then adding the value of each number in the list to S. S would first take the value 1, then 3, then 6, and, last, 10, which would be returned as the result. This process can be visualized with Figure 1 (left panel). It results in acquisition of Chunk C1:

> **IF** the task is to compute-sum of [1,2,3,4]
> **THEN** respond "sum is 10"

The action or right-hand side of Chunk C1 is to return the result, "Sum is 10." The conditions or left-hand side of Chunk C1 consists of all of the items in the initial state that were required to find the result, which in this case means the function name and the numbers that were summed.

The chunking mechanism models a person who encodes an episodic memory of his or her experience computing sums. The acquisition of Chunk C1 buys the problem solver an efficiency advantage. Chunk C1 is stored in parallel recognition memory and on future occasions will propose its answer as soon as its conditions are met, thereby obviating the need to recompute a result by summing the individual numbers in the list. This process is illustrated with Figure 1 (right panel): The calculation needed in Figure 1 (left panel) has been replaced by the firing of Chunk C1. In the terminology of explanation-based learning (Mitchell, 1986), the chunk is an *operationalized* form of the knowledge of how to compute sums: It improves the efficiency of the problem solver.

The conditions of Chunk C1 illustrate the consequences of Soar's backtracing mechanism. In Section 3.1, we see that backtracing has what may at first appear to be surprising consequences. Backtracing ensures that the conditions of a chunk that returns a result R consist of all of the items used to form R. In this article, we consider a "strict" version of Soar in which backtracing cannot be avoided.[2] We believe that the arguments advocated by Vera, Lewis, and Lerch (1993) and expanded in the present article support the view that "strict" backtracing should not be avoided in the Soar architecture.

3.1. Users Acquire Display-Based Skills

Larkin (1989) coined the term *display based* to refer to skills that are inherently dependent on cues from the environment. Larkin described how, when making coffee, people need not remember the sequence of goals. Instead, the state of the external world cues which tasks can be done when (e.g., water cannot be poured into a mug until the mug has been retrieved from the cupboard).

There has been much interest in studying display-based skill within HCI (Altmann, Larkin, & John, 1995; Bauer & John, 1995; Briggs, 1990; Draper, 1986; Franzke, 1995; Howes, 1994; Howes & Payne, 1990; Howes & Young, 1996; John, Vera, & Newell, 1994; Kitajima, 1989; Kitajima & Polson, 1995; Mayes, Draper, McGregor, & Oatley, 1988; Payne, 1991; Vera et al., 1993). An important result in this literature is that, under certain circumstances, expert users of computer systems may be incapable of recalling how to perform a task (when away from the device) even though they do it regularly in their everyday work (Mayes et al, 1988; Payne, 1991).

Vera et al.'s (1993) model of display-based skill offers a Soar account of these results. Their model of instruction taking learns how to get cash from an automatic teller machine (ATM) and shows how the backtracing mechanism necessarily leads Soar to acquire display-based chunks for

---

2. In the Soar version current when this article was written, backtracing could be avoided by using a particular type of preference.

interactive tasks (i.e. tasks for which actions are directly cued by the device). That is, chunk conditions refer to the state of the display. Applied to the task of opening a file in Microsoft Word, Vera et al.'s analysis yields Chunk C2:

> **IF** the task involves using a file
>   & the item "file" is on the display
> **THEN** move the mouse pointer to the item "file"

Chunk C2 is display based (i.e., the *file* item must be present on the display for it to fire). Why do we claim that Soar must learn chunks (e.g., C2) with display-based conditions? We consider three examples—learning from instruction, learning from exploration, and comprehending a computer program. We show that the claim follows in each case.

Learning From Instruction

If Soar is given the instruction, "Move the mouse to the file option," then it can determine that the action required is to move the mouse pointer to the *file* item on the display, and it can learn a rule that has this action as its right-hand side.[3] But, what will be the conditions of the learned rule? The backtracing mechanism of Soar determines rule conditions by grouping the information that contributed to the proposal of the action. In the compute-sum example, these conditions are the function name and the numbers that are summed. They can be thought of as the "reasons" that an answer of 10 is proposed. However, in the "Move the mouse to the file option" example, the reasons that Soar proposes to move the mouse to the *file* item are because the model is instructed that this is the right thing to do and because the model successfully provides an interpretation, or rationale, for the instruction in terms of the task. Therefore, rather than learning Chunk C2, Soar learns Chunk C3:

> **IF** the task involves using a file
>   & there is an instruction "move the mouse to the file option"
> **THEN** move the mouse pointer to the item "file"

Chunk C3 is not what is required, and it raises a problem. In fact, Chunk C3 appears to be useless, as it will not recall the required action ("Move the mouse pointer to the file option") given just the task description and the display. Instead, it will fire only when an instruction is

---

3. In general, a semantic match process is required to determine the appropriate label.

present. The conditions seem to imply that, despite the fact that Soar has been instructed how to do the task, it will continue to require instruction on subsequent trials.

Chunk C3 is instruction based. Similar chunks form whenever instruction is used to determine the right-hand side of a rule. The apparent uselessness of instruction-based chunks seems to suggest that, although chunking is good at learning the results of computations that the problem solver already knows how to perform (e.g., sums), it cannot learn new rules that do not derive from knowledge the system already has. This problem, termed the *data-chunking problem,* was first observed by Rosenbloom, Laird, and Newell (1987).

Fortunately, Rosenbloom and Aasman (1990), Rosenbloom et al. (1987), Rosenbloom, Laird, and Newell (1988), and Vera et al. (1993) have offered a way to use instruction-based chunks when there is no instruction available. The first part of the solution is to realize that these chunks can be used to *recognize* instructions that were given previously. When an instruction such as "Move-to file" is given for the first time, a Soar model must problem-solve in order to understand the instruction, and instruction-based Chunk C3 is formed as a consequence. But, when the model is given the instruction again, Chunk C3 fires, obviating the need to redo the interpretation. That Chunk C3 fires can be used by Soar to infer that the "Move-to file" instruction was given previously (i.e., to recognize the instruction). The second part of the solution is to propose that, in the absence of external instruction, the way to recall what was instructed is to *reconstruct* the instruction internally (i.e., guess what the instruction was) and then hope that one of the instruction-based chunks fires, thereby "recognizing" which instruction was given previously. This *generate-and-recognize* model of memory retrieval was proposed many years ago in the psychology literature (Anderson & Bower, 1972; Kintsch, 1970).

The question then is: How do computer users construct appropriate guesses? Several authors have reported models that use the external display as a generator of possible actions (Altmann et al., 1995; Bauer & John, 1995; Howes & Payne, 1990; Howes & Young, 1996; Kitajima & Polson, 1995). Vera et al. (1993) tied this idea into Soar as a solution to the reconstruction of instructions. Vera et al. suggested that display items are examined until an instruction-based chunk fires. Soar must examine each display item in turn (e.g., *Apple, File, Edit,* ...) and internally "imagine" that an instruction has been given to select that item. Display items continue to be examined until the conditions of some instruction-based chunk are met, whereupon the chunk fires, recognizing the display item. In this way, the device display acts as a "reminder" of the content of the instructions.

For example, after Soar learns a set of instruction-based chunks for a task, when it starts that task again, there are two possible courses of

Figure 2. Learning to recall an instructed action by generating and recognizing candidate actions. Left panel: C3 recognizes the action; Right panel: C2 recalls the action. *Note.* From "Learning consistent, interactive and meaningful device methods: A computational model," by A. Howes & R. M. Young, 1996, *Cognitive Science, 20,* p. 318. Copyright © 1996 by Cognitive Science Society, Inc. Reprinted with permission.

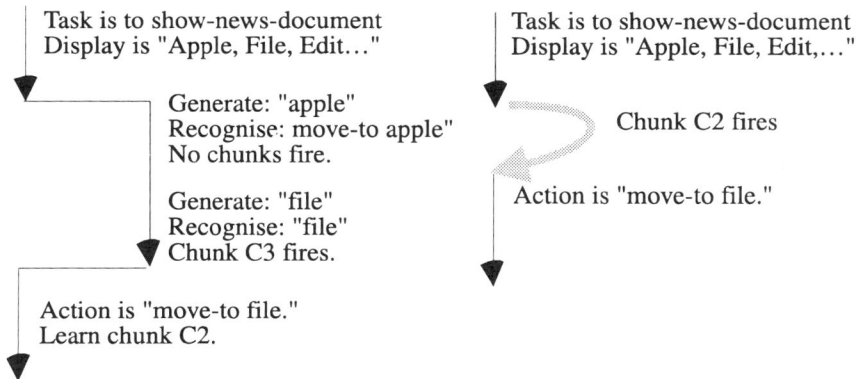

Task is to show-news-document
Display is "Apple, File, Edit..."

Generate: "apple"
Recognise: move-to apple"
No chunks fire.

Generate: "file"
Recognise: "file"
Chunk C3 fires.

Action is "move-to file."
Learn chunk C2.

Task is to show-news-document
Display is "Apple, File, Edit,..."

Chunk C2 fires

Action is "move-to file."

action—either ask for instruction again or attempt to remember the instruction by generating plausible options and recognizing one. Suppose the model chooses to generate and recognize and that it selects as generator the labels on the display. First, it examines the *apple* menu item and constructs the instruction, "Move-to apple," but no instruction-based chunk fires. Then it examines the word *file* and constructs the instruction, "Move-to file." For this instruction, the conditions of Chunk C3 are met, so the chunk fires and proposes the action of moving the mouse to the *file* item. The required action is thereby recalled by the use of a generate-and-recognize algorithm (Figure 2, left panel). In this process, a new chunk is learned that operationalizes the problem solving that found the action. On this occasion, there was no external instruction, only an internally imagined one. What did contribute to the formation of the chunk was the display and the task. Hence, the new rule (Chunk C2) does not require the instruction to be present in order for it to fire (Figure 2, right panel), but it is display based.

To summarize this first example, we see that the way in which Soar's backtracing mechanism determines chunk conditions means that it must use a generate-and-recognize method in order to recall instructions. In cases in which the external display is used as a generator, this process necessarily leads to display-based chunks, and the architecture therefore determines that Soar models capture the data of Mayes et al. (1988) and Payne (1991).

Learning From Exploration

Soar's backtracing mechanism also constrains the exploratory acquisition of knowledge about the sequence of actions to achieve a goal. Ayn (Howes, 1994) is a model (implemented in Prolog and Soar)[4] that learns menu structures by acquiring recognition chunks that encode which options the model has selected and which lead to the goal. For example, when Ayn finds a goal state after some exploration, it is then in a position to learn that the previous action leads to success. Again, all of the conditions of the chunk are external to the problem solving, and Chunk C4 is recognitional:

> **IF** the previous action was to select "open"
> & the current state is a goal state
> **THEN** the action "open" is correct.

In order to use that "outcome-based" chunk to guide its behavior toward the goal state, Ayn must use a generate-and-recognize technique analogous to that used for instruction learning. Using the options available in the previous state as a source, it must generate each action and "imagine" it leading to the goal state. For the correct choice, a recognition chunk will fire. Thus, does the option *New* lead to success? No recognition chunk fires. Does *Close* lead to success? No chunk fires. Does *Open* lead to success? Chunk C4 fires, so *Open* must be the correct choice. The result of that problem solving is a new chunk recalling that *Open* leads to the goal state, but it is still display based insofar as one of its conditions is that the *Open* option be available on the screen.

To summarize the first two examples, if the display is the only means for a Soar program to generate possible actions, then the acquired skill will be display based. Further, this appears to be true for both learning from instruction and learning from exploration. Thus, through its chunking mechanism, the Soar architecture captures the HCI data indicating that expert performance is display based (Mayes et al., 1988; Payne, 1991). The architecture achieves this by virtue of its backtracing mechanism for determining chunk conditions, which requires the use of the generate-and-recognize process for the acquisition of recall knowledge.

Comprehending a Computer Program

Altmann (1996) and colleagues (Altmann et al., 1995) described a Soar model of a programmer's scrolling behavior during comprehension of a

---

4. The Soar implementation of Ayn was done in collaboration with John Rieman.

computer program. One of the aspects of human behavior captured by the model concerns circumstances in which the programmer uses scrolling in order to display hidden information. Consider, for example, a situation in which, while comprehending a Pascal program, the model comes across a definition for a function called calculate_cost.[5] The model attempts to comprehend what this function will achieve and, in so doing, encodes recognition chunks in exactly the same way as the models in the previous examples.

Now imagine that, at a later time, after the definition of calculate_cost is no longer on the display, the model encounters a call to the function. In this situation, the model again needs to know what calculate_cost does, and it again sets the goal of comprehending the function. In order to comprehend the function, the model must use the definition of calculate_cost, which, due to the backtracing constraint, it has not encoded in a recallable form. The model does not know where in the program file the definition of calculate_cost is located. However, if it recognizes having previously comprehended the function, then it can use this knowledge to direct scrolling to parts of the program file that have already been seen. For this reason, the model invokes a scroll action until the required function is displayed.

In this way, Altmann et al.'s (1995) model captures the fact that, when trying to understand a program, programmers do not simply read through the file line by line; instead, they jump around examining and reexamining parts of the code, as they are needed, in a nonlinear fashion. Due to Soar's backtracing mechanism, this behavior is exactly what emerges in Altmann et al.'s model.

3.2. Recapitulation: Recognition, Reconstruction, and Recall in Soar

At this point, it might be helpful to summarize what we have discovered from the analysis of the relation of Soar's chunking mechanism to display-based skills, because our discussion of the other empirical regularities we examine make repeated reference to these basic results. We have worked through the following argument:

- A Soar model cannot directly acquire rules that recall information originating in the external world. The desired information can be captured on the right-hand side of a chunk, but, because the rationale for this right-hand side is derived from the environment, the information will be included on the left-hand side too. The chunk is

---

5. For clarity, we have assumed a Pascal program in which Altmann et al.'s (1995) original model dealt with a Soar program.

therefore "circular" (e.g., instruction-based chunks are conditional on the presence of the instruction and are of little use at first glance). Importantly, these chunks arise automatically from problem solving (e.g., the process of interpreting an instruction).

- Such circular chunks encode recognition knowledge. Given the instruction (and assuming that the other conditions are met), the instruction-based chunk will fire, in effect telling the Soar model "Yes, I have seen that instruction before."
- In order to acquire a recall chunk that gives the Soar model information without that information present, the model must first reconstruct the information. It does so by generating suitable candidates and using the circular recognition chunks to identify which candidate is correct.
- In many HCI settings, the most easily available source for the generator—or the only source—is the external display. Recall chunks formed from such a generator encode a display-based skill, because the chunks depend on the external presence of the cue for action (e.g., an item to be selected).

We shall see that some HCI situations demand the use of an internal generator, because no external source is available. In the analysis of the remaining HCI phenomena, much of the discussion focuses on this question of the ease or difficulty of generating the plausible candidates.

3.3. Keyboard Users Need Meaningful and Mnemonic Command Names

The role of meaningfulness and mnemonics in the design of keyboard-based devices has been studied over many years (e.g., Barnard & Grudin, 1988; Furnas, Landauer, Gomez, & Dumais, 1987). It is a truism that a good mnemonic helps recall. Consider the user of a keyboard-based device who, having been told the name of the command that gets rid of a file, needs to recall it. What was the command? Was it some arbitrary letter sequence, *CTRL–X* perhaps? Or was it some arbitrary word, perhaps *rabbit?* Alternatively, the command name may reflect the semantics of the task. It may be something derived from "get rid of" (e.g., *delete, remove*) or some other word with a meaning related to the task. It may also be an abbreviation of one of these words. Data support the view that meaningful and mnemonic command names help recall. Can Soar's backtracing mechanism capture this effect?

In Unix, the command that gets rid of files is *rm,* derived from the word *remove.* As in Section 3.1, in order to learn this command from instruction, a Soar model must first acquire instruction-based chunks, as in Chunks C5:

IF the task involves getting rid of something
    & there is an instruction to use the word "remove"
THEN use the word "remove"

and C6:

IF the task involves using the word "remove"
    & there is an instruction to use the first two consonants
THEN use the first two consonants

As we have seen, when given the "get rid of" task again, the only way that a Soar model can recall the instructions it was given is by a generate-and-recognize process. If we assume that the model has the semantic knowledge to generate appropriate words from the "get rid of" task description, then an internal generator based on processes of semantic and lexical search can generate plausible words such as *delete, cut, throw out,* and eventually *remove.* (The semantic knowledge may be derived from a technique such as that suggested in Miller & Charles, 1991, and programmed into Soar's production memory.) After *remove* is generated, Chunk C5 will fire, indicating that *remove* was the instructed word. Subsequently, a second generate-and-recognize process—this time generating types of abbreviation (first two letters, first letter, ..., first two consonants)—should lead Chunk C6 to fire.

Now contrast the case of *rm* to another Unix command, *grep* (derived from *get regular expression*), which is used for finding files with particular contents. Given the instructions for this command, a Soar model would learn instruction-based Chunks C7:

IF the task involves finding a file by contents
    & there is an instruction to use the word "get regular expression"
THEN use the word "get regular expression"

and C8:

IF the task involves using the word "get regular expression"
    & there is an instruction to take the first letter
        of the first word, the first two letters of the second ...
THEN take the first letter of the first word, the first two letters of the second ...

Chunks C7 and C8 encode the instructions, but they can be used only for recognition. In order to extract the information implicit within them, a generate-and-recognize process must be used. But on what basis can plausible candidates be generated? It would be difficult to justify a Soar model that, given the task to find a file by contents, just happened to generate the phrase *get regular expression*. Although this may be an accurate model for some, most users who do not already know the command are unlikely to generate it spontaneously. Instead, the model must include back-up generators that operate when semantic generators fail. What the back-up generators are and how they work are less clear. It may even be the case that an elaborated set of recognition rules would have to be acquired to guide the generator. Perhaps separate recognition rules are learned for each word, other rules recognize the first letters of the word, and still other rules recognize syntactic types. Perhaps, as seems more likely, people learn *grep* as a new word rather than decomposing it as an acronym/abbreviation. The details are beyond the scope of this article, but it is clear that the generate-and-recognize process would be less constrained for *grep* than for *rm* and, because the search space is potentially much larger, more time-consuming and prone to failure.

In summary, because a Soar model capable of learning keyboard commands from instruction requires the use of semantic generators, computer interfaces designed with meaningful and mnemonic command names will be easier to learn than those employing arbitrary command names. This observation indicates that Soar is aligned with broad aspects of the psychological data on command naming.

3.4. Display-Based Devices Are Easier to Learn Than Keyboard-Based Devices

Display-based devices constrain the next action that the user selects by forcing a choice from a limited set of externally presented options. In contrast, keyboard-based devices require the user to recall sequences of command words and to enter them. This constraint gives display-based devices a considerable advantage over keyboard-based devices for exploratory learning. This advantage was demonstrated by Charney, Reder, and Kusbit (1990) and is illustrated by the fact that the overwhelming majority of studies of exploratory learning in HCI is carried out using display-based devices (Carroll & Rosson, 1987; Franzke, 1995; Lewis, 1988; Polson & Lewis, 1990; Rieman, 1994; Robert, 1987). The difference in learnability between display-based and keyboard-based devices under exploratory learning conditions is easy to understand without recourse to the architectural learning mechanism. In one case, actions are generated from the external world; in the other case, they must be generated from the user's knowledge base. If users have had no prior exposure to a keyboard-based device, there is no reason to suppose that they can generate the correct action without help (see Section 3.3). In fact, Furnas et al. (1987) showed that even the most careful design can achieve only low guessability of command names without external support.

But, is there a difference in learnability between keyboard-based and display-based even when users are instructed? Under instructional learning conditions, users can be given sufficient and informationally equivalent advice (see Larkin & Simon, 1987, for a definition) on how to use either kind of device. However, that users can be given instructions does not necessarily mean that they can learn them. As we have seen, users cannot learn to retrieve arbitrarily complex instructions at will. One of the factors affecting the likelihood that instructions will be retrieved is the cues provided by the display of the device being learned. Does the backtracing mechanism of Soar predict a difference in learnability, or can an analyst program Soar with model increments that find either kind of interface equally easy or difficult to learn?

Again, in order to acquire rules that directly recall instructions, a Soar model must be programmed with a generate-and-recognize algorithm. Howes and Young (1996) reported on a Soar model called *task-action learner* (TAL), which uses the generate-and-recognize process to learn both Microsoft Word tasks and Unix tasks. TAL uses the display as a source of candidate instructions for the display-based device (see Section 3.1). But, to use a keyboard-based device, TAL must use four internal generators in a hierarchical fashion. The model works downward from the task description, decomposing it into finer grain subtasks until an action that can be performed on the device is found. First, it determines which feature of the task is to be communicated to the device (e.g., effect = "get rid of"); second, it determines which word is to be used in order to communicate the feature (e.g., *remove*); third, it determines whether and how to abbreviate the word (e.g., take first two consonants); fourth, it computes the abbreviation (e.g., *rm*) and types the characters on the keyboard. The four generators use long-term memory as their source of candidate actions, each attempting to reconstruct one of the steps.

The internal generators define a search space in the same way that the menu labels and icons on a display-based device do. In both cases, the instruction-based chunks can be used as recognitional control knowledge to help the problem solver find the instructed path through this space. For HCI, the important differences are that (a) the internal search space required for keyboard-based devices is almost invariably larger than the external search space required for display-based devices; (b) the internal search space is subject to differences in individuals' knowledge in a way that the external search space is not (see Furnas et al., 1987); and, (c) with keyboard-based devices, Soar (like people) must rely on mnemonic encoding of commands in order to make the process of reconstructing instructions reasonably efficient (see Section 3.3).

In consequence, we can say that the backtracing mechanism forces Soar models to exhibit a learnability difference between display-based and keyboard-based devices. The difference stems from the architecture, not from what the analyst chooses to put in the model increments.

3.5. Consistent Interfaces Are Easier to Learn and Use

There are many ways in which a computer interface can be consistent. Our primary interest here is in the internal consistency of device methods—that is, the consistency of methods or of task-action mappings (Payne & Green, 1986) with each other for a single device. The internal consistency of a device constrains learning when users assume that task-action mappings learned for one task will apply to all semantically similar tasks. For example, if a method for opening a file has a certain syntax, then people will assume that the mapping for closing a file will have the same syntax.

Consider syntactic consistency and mnemonic consistency. An interface is syntactically consistent if the order in which types of items occur is the same across a broad set of tasks. For example, if, in the command to copy a file, the word *copy* must be typed before the description of the file, then the user may expect that, in the command to delete a file, the word *delete* must also be typed first. The mnemonics of a language are consistent if the way in which abbreviations are derived from command names is the same across many tasks. For example, if the command to *remove* a file is executed by typing the first two consonants of the command name, then it will help learnability if other commands are executed also by taking the first two consonants of their names.

Payne and Green (1986, 1989) reported several experiments to show that consistent interfaces are easier for people to learn and use. For example, Payne and Green (1989) found that the number of syntax errors made by subjects was greater for devices with inconsistent grammars. Other studies (e.g., Barnard, Hammond, Morton, Long, & Clark, 1981; Kellogg, 1987; Lee, Foltz, & Polson, 1994; Payne, 1985) support these findings. Can the backtracing mechanism of the Soar architecture predict these results, or is it the case that a model constructed in Soar can be programmed to find either a consistent interface or an inconsistent interface easier to learn?

Howes and Young's (1996) TAL (see Section 3.4) uses assumptions from the task-action grammar (TAG) theory of command-language consistency (Payne & Green, 1986, 1989). Tasks in TAG are represented as feature-value pairs. For example, a task to open a file might be described by [Effect=open, Object=file, Name=newsletter]. This representation provides TAL with the flexibility to learn chunks that are sensitive only to the feature name and not to the feature value. For example, Chunk C9 encodes the knowledge that, for all tasks that contain an Effect feature, the first subtask should be to communicate the value of this feature to the device:

IF task contains the feature Effect
    & there was no previous subtask
THEN make the subtask to communicate Effect

These generalized chunks align TAL with Payne and Green's (1986, 1989) data. TAL learns consistent interfaces faster than inconsistent ones because the generalized chunks mean that less instruction is required, and it makes errors on inconsistent interfaces because of the now overgeneral chunks.

However, the generalized chunks are acquired only because of the way that Howes and Young (1996) programmed TAL to generalize across feature values. There appears to be no architectural constraint forcing models to be that way. A Soar model could just as easily have been programmed to learn rules of the form (Chunk C10):

IF task contains the feature Effect = open
    & there was no previous subtask
THEN make the subtask to communicate Effect = open

A model using these specific chunks would learn a consistent interface as slowly as an inconsistent one, because, in either case, it would have to acquire an independent rule for each value of the feature, and it would fail to make errors that are due to overgeneralization.

A tentative[6] conclusion must be that Soar's backtracing mechanism does not force the learning of consistent interfaces to be easier than that of inconsistent ones. Unlike for the other HCI phenomena we have examined so far, a Soar model that makes predictions about the effects of interface consistency in line with the empirical data does so not because of anything inherent to Soar but by virtue of the way the model increment has been programmed.

3.6. Learning Locational Knowledge

An important aspect of skill for display-based devices is knowledge about where items are located on the display (see, e.g., Lansdale, 1991). A user who knows that the "wastebasket" is located in the bottom right of the display need not spend time looking for it. Cognitive psychology has seen a long-running debate about whether locational knowledge is acquired

---

6. We use the word *tentative* because the analysis here is only shallow. It takes no account, for example, of how the knowledge about what aspects of the task to attend to is itself acquired. A deeper analysis might perhaps reveal ways in which Soar is biased toward learning the generalized rules—or it might not. The topic lies beyond the scope both of this article and of our current understanding.

automatically. For example, Andrade and Meudell (1993) claimed that their data show automatic memory for locations, and Naveh-Benjamin (1987) claimed that spatial tasks fall along an ease-of-learning continuum; Lansdale (1995) claimed that there is little evidence either way.

Does the Soar architecture force Soar models to be aligned with either side of the debate? Operationalizing the notion of automaticity for a cognitive architecture such as Soar is difficult. However, here we argue that Soar will not automatically learn to recall locational information. Instead, a Soar model would have to make a deliberate effort to learn locations.

In an unreported Soar model programmed by co-author Howes, the task involves "visually" finding the right menu item by its label (e.g., *Format*) on a simulated menu and then moving the mouse to the location of that label. Chunks such as C11 are formed:

> **IF** the task is to move the mouse to "Format"
> & "Format" is in the middle
> **THEN** move the mouse to the middle

Chunk C11 adds knowledge that recognizes the location of *Format*. It can be learned only after the location has been determined externally. Acquisition of these recognition chunks is automatic and a direct consequence of the application of the backtracing mechanism to the current problem-solving task.

As before, because locational knowledge is acquired initially within recognition chunks, a generate-and-recognize process is needed for the knowledge to be captured in a form in which it can be retrieved and used to guide behavior. To implement a generate-and-recognize process for the target *Format,* the problem solver might pose the questions "Is *Format* to the right? Is *Format* to the left?" and continue doing so until the options are exhausted or until a chunk such as C11 fires. However, whether a model engages in this generate-and-recognize process is a matter of strategic choice. Because the item to be found (e.g., *Format*) is available on the device display, the location need not be retrieved from memory for the task to be achieved. It may be the case that scanning for the item is a more efficient way of achieving the task on this trial than taking the time to generate and recognize. Thus, although learning to recall a location would have a long-term benefit (i.e., on trials subsequent to its acquisition, it will decrease performance time), on the trial in which it is acquired, the process of acquiring the chunk may increase performance time for the overall task.

In the case described so far, knowledge used to guide the strategic choice is part of the model increment, which of course may itself be dependent on knowledge of the task environment. In other cases, the task environment dictates that locational knowledge be acquired. Bauer and

John (1995) described a Soar model of a player of a Nintendo video game. In order to do well at the game, the player must learn to avoid enemies by jumping over them. To jump successfully, the distance between the player and the enemy must be exactly right. If the distance is too short, the player will not have time to get off the ground; if the distance is too long, the player will land before the enemy has passed underneath. The only way for the Soar model to learn the optimal distance is, as before, by first learning recognition chunks and then using the generate-and-recognize algorithm to acquire recall chunks. The recognition chunks encode the distance of the enemy when the jump is launched and the result of the jump (success or failure). Generation involves "imagining" various distances and outcomes and choosing behavior in the current context according to what is recognized. As in the case of learning to recall menu locations, the architecture determines that to learn to recall jump locations, the generate-and-recognize algorithm is required. However, the two examples differ in that the task environment determines that, if the Nintendo locations are not learned, then the player will be "killed," and the task will not be completed. Another Soar model that learns to operate in a task environment in which locational knowledge is essential was described by Miller, Lehman, and Koedinger (1997).

To summarize this discussion of locational knowledge, we can say that, as a consequence of the backtracing mechanism, the Soar architecture may acquire recognition chunks of locations automatically, but whether recall chunks are subsequently learned from these recognition chunks depends on whether the problem solver decides to take the time to go through the generate-and-recognize process. This choice is determined not by the architecture but by the task environment or by the way in which the model trades off immediate performance time against benefit of learning.

4. DISCUSSION

4.1. Constraints From Soar's Learning Mechanism

In Section 3, we examine in some detail the ways in which Soar's chunking mechanism determines aspects of the behavior of Soar models that learn. We have seen examples of "hard" constraints, in which the cognitive architecture necessarily enforces certain properties on the model. We saw, for example, how Soar models that learn to operate a certain class of interactive device will necessarily acquire a display-based skill, whether they learn by instruction or exploration. The backtracing mechanism of the Soar architecture leads automatically to the acquisition of recognition chunks, which in turn leads to the need for a generate-and-recognize mechanism in order to learn recall chunks by reconstructing the context of the recognition chunks. This mechanism in turn determines that a Soar model of a user must necessarily acquire display-based skills. Similarly, we have seen how, given certain plausible assumptions about a device, a display-based interface is necessarily easier for a Soar model to learn than is a keyboard-based interface. Again, the nature of Soar's chunking mechanism forces the result that learning by instruction will be easier for meaningful or mnemonic command names, which allow users to reconstruct instructions for keyboard-based devices.

The generate-and-recognize model of retrieval has been established for many years in cognitive psychology (e.g., Bahrick, 1970). The similarity between Soar's emergent technique and other generate-and-recognize models (e.g., Anderson & Bower, 1972; Kintsch, 1970) was observed by Rosenbloom et al. (1987). Our arguments here add several HCI phenomena to the list of those explainable in terms of a generate-and-recognize model. Further, that the generate-and-recognize model is a consequence of the Soar architecture adds weight to the claim that Soar's learning mechanism has psychological plausibility.

In contrast to the strong role attributable to Soar's backtracing mechanism in accounting for the phenomena just mentioned, other empirically observed regularities seem to depend on the assumptions built into particular Soar models. For example, the TAL model requires less instruction and makes fewer errors when learning consistent as opposed to inconsistent interfaces, because the model is programmed with an algorithm that learns recognition rules generalized according to the semantics of the task, and the evidence is that people do the same. However, Soar could just as easily be programmed with an algorithm that acquires specific recognition rules that do not generalize to whole categories of task. We could, for example, construct a Soar model in which the recognition rule for an instruction to use the *cut* menu option to delete the word *rabbit* contains a reference to the word *rabbit*. The model would then not make the generalization that it should use *cut* for deleting other words too.

The topic of the initial acquisition of recognition rules does seem to be an area in which the Soar architecture is currently underconstrained. As it stands, Soar can acquire in one learning event a recognitional chunk of arbitrary complexity and will never confuse it with another chunk no matter how similar their conditions We believe that a key research objective for people working with the Soar architecture should be the development of constraints on the acquisition of recognition chunks—to reflect the fact that people's ability to discriminate between objects is dependent on aspects of distinctiveness, familiarity, and frequency. Some Soar techniques (e.g., Miller & Laird, 1992) may offer solutions to this problem, as may perhaps other approaches based on ideas of discrimination networks (Richman, Staszewski, & Simon, 1995), but at present they are not an integral part of Soar.

Last, although it has not been a major theme in this article, we have also seen how some aspects of a model's behavior are determined by the task environment. This point is by now so obvious that we will not labor it (see, e.g., Vera et al., 1993). It is, however, worth restating the role of the environment in determining strategic choices such as whether to make the effort to learn locational knowledge: In some environments, locational knowledge is essential to the task; in others it merely provides a gain of efficiency.

4.2. Role of the Architecture in Soar Models

We have seen that in the case of "hard" constraints, the Soar architecture enforces certain necessary properties on a model. For the properties discussed earlier, the backtracing mechanism of learning enforces these results: The architecture cannot be programmed with algorithms that avoid these behavioral outcomes. In such cases, the existence of the hard constraint makes assigning responsibility for the phenomenon to the cognitive architecture straightforward. It becomes meaningful and correct to claim, for example, that Soar predicts that learning to use an interactive device through exploration will result in a display-dependent skill. We are able to go beyond an assertion such as, "Within Soar, the modeler has the possibility of constructing models of learning that result in display-dependent skill," which implies that the modeler has a choice and that Soar would allow the construction of models that do not share that property.

Conversely, we have seen cases in which a feature of the model's behavior is not dictated by the architecture. For example, it appears that taking advantage of consistency in the interface to learn generalized rules is something that depends on the strategic knowledge in the model increment and about which the architecture itself is neutral.

However, the responsibility of the architecture in shaping the behavior of models is not always the black-and-white issue that these examples might suggest. It is not always the case that the architecture either enforces a property or is totally neutral about it. In Section 2.2, we examined soft constraints, in which the architecture "favors" a property without actually enforcing it. We were led to the notion of compliancy to express the degree to which the model increment allows the architecture its say in determining behavior.

The issue of compliancy is central to the questions addressed in this article—questions about the consequences of constructing user models within a cognitive architecture and about how to allocate responsibility for the predictions made about users' behavior between the architecture itself and the particular model increment. With models of a moderate to high degree of compliancy, a good part of the credit or blame lies with the architecture, which is playing a moderate- to high-profile role in determining the predictions of behavior. In the extreme case, as we have seen with some of the models that exploit Soar's learning mechanism, the architecture may be imposing a "hard" constraint on the model, so that certain aspects of the predictions are the clear responsibility of the architecture. With low-compliant or noncompliant models, any guidance offered by the architecture is being disregarded or overruled by the model increment, with the result that the predictions of behavior are due almost entirely to the model increment (i.e., to the way the model has been programmed).

5. CONCLUSIONS

This article began by noting that a style of user modeling, in which models are constructed within a fixed cognitive architecture, raises novel problems of evaluation for both practical and scientific purposes. We asked about how the responsibility for any empirical successes or failures of such user models can be apportioned to the architecture within which the model is built or to the model increment that specifies the particular model.

We saw that, in some cases, it may be possible to give a clear-cut answer to the question. If the architecture imposes a "hard" constraint on the model, so that any model built within the architecture exhibits a certain feature (e.g., display-based skill), then we can clearly point to the architecture as being responsible for that feature. Conversely, if the architecture is neutral on some point, allowing models to be built that equally well display one feature or another (e.g., a learning advantage for consistent interfaces), then we can unambiguously attribute the feature to the model increment (i.e., to the way the model has been "programmed"). We also saw that the question does not always permit such black-and-white answers. The behavior of the model, after all, depends in general on both the cognitive architecture and the particular model specification (as well as on the task environment), so the responsibility for the behavior of the model must in some sense belong jointly to both.

We can approach the issue of joint responsibility from either side. Coming from the architecture side, as it were, we meet the notions of hard and soft constraints, in which a hard constraint enforces a property on the model, whereas a soft constraint merely makes it easier to build the model with a particular property than not. The notion of soft constraints is obviously graded, as there can be a greater or lesser difference in the ease of building the model one way or the other. Coming from the model side, we encounter the idea of compliancy. A compliant model allows the architecture to play a role (when possible) in guiding the model's behavior. A noncompliant model blocks the architecture's role either by disregarding it or by opposing it. Again, compliancy is obviously a graded concept. Models can be more or less compliant with their architecture. (We also note that a model can be more or less compliant with the guidance being offered by the environment.)

We have seen also that, at least for Soar, we can concretize these ideas about compliancy. Soar has built-in mechanisms for the selection and ordering of processing steps, although these can of course be overridden by additional control knowledge. A Soar model is compliant to the extent that it allows processing to follow the course proposed by these mechanisms. It is noncompliant to the extent that it overrides them.

Although doubts are sometimes expressed as to whether cognitive architectures have any empirical content and therefore any scientific contribution to make to user modeling, our analysis shows that indeed they do. Architectures play their part by imposing theoretical constraints on the models constructed within them. In some cases, such constraints can be strong enough to dictate that models exhibit certain characteristic properties. In other cases, the architecture can influence the model more subtly: The extent to which it is allowed to do so depends on an aspect of the way a particular model is specified—its compliancy.

NOTES

Acknowledgments. Thanks to Bonnie John, Wayne D. Gray, Erik Altmann, and two anonymous reviewers for helpful comments on a draft of this article.

Support. The work reported in this article was done as part of a project funded by the UK Joint Councils Initiative in Cognitive Science and HCI.

Authors' Present Addresses. Andrew Howes, School of Psychology, Cardiff University of Wales, P.O. Box 901, Cardiff, CF1 3YG, United Kingdom. E-mail: howesa@ cardiff.ac.uk. Richard M. Young, Department of Psychology, University of Hertfordshire, Hatfield, Herts, AL10 9AB, United Kingdom. E-mail: r.m.young@herts.ac.uk.

HCI Editorial Record. First manuscript received February 1996. Revision received November 1996. Accepted by Wayne D. Gray. Final manuscript received March 25, 1997. — *Editor*

REFERENCES

Altmann, E. M. (1996). *Episodic memory for external information* (CMU–CS–96–167). Doctoral dissertation, School of Computer Science, Carnegie Mellon University, Pittsburgh.

Altmann, E. M., Larkin, J. H., & John, B. E. (1995). Display navigation by an expert programmer: A preliminary model of memory. *Proceedings of the CHI'95 Conference on Human Factors in Computing Systems,* 3–10. New York: ACM.

Anderson, J. R. (1993). *Rules of the mind.* Hillsdale, NJ: Lawrence Erlbaum Associates, Inc.

Anderson, J. R., & Bower, G. H. (1972). Recognition and retrieval processes in free recall. *Psychological Review, 79,* 97–123.

Anderson, J. R., Matessa, M., & Lebiere, C. (1997). ACT–R: A theory of higher level cognition and its relation to visual attention. *Human–Computer Interaction, 12,* 439–462. [this special issue]

Andrade, J., & Meudell, P. (1993). Is spatial information encoded automatically in memory? *Quarterly Journal of Experimental Psychology: Section A—Human Experimental Psychology, 46,* 365–375.

Bahrick, H. P. (1970). Two-phase model for prompted recall. *Psychological Review, 77,* 215–222.

Barnard, P. J. (1987). Cognitive resources and the learning of human–computer dialogues. In J. M. Carroll (Ed.) *Interfacing thought: Cognitive aspects of human–computer interaction* (pp. 112–158). Cambridge, MA: MIT Press.

Barnard, P. J., & Grudin, J. (1988). Command names. In M. Helander (Ed.), *Handbook of human–computer interaction* (pp. 237–255). New York: Elsevier.

Barnard, P. J., Hammond, N., Morton, J., Long, J. B., & Clark, I. A. (1981). Consistency and compatibility in human–computer dialog. *International Journal of Man–Machine Studies, 15,* 87–134.

Barnard, P. J., & May, J. (1993). Cognitive modelling for user requirements. In P. Byerley, P. J. Barnard, & J. May (Eds.), *Computers, communication and usability: Design issues, research and methods for integrated services* (pp. 101–146). Amsterdam: North-Holland.

Barnard, P. J., Wilson, M., & MacLean, A. (1988). Approximate modelling of cognitive activity with an expert system: A theory based strategy for developing an interactive design tool. *Computer Journal, 31,* 445–456.

Bauer, M. I., & John, B. E. (1995). Modeling time-constrained learning in a highly-interactive task. *Proceedings of the CHI'95 Conference on Human Factors in Computing Systems,* 19–26. New York: ACM.

Briggs, P. (1990). Do they know what they're doing? An evaluation of word-processor users' implicit and explicit task-relevant knowledge, and its role in self-directed learning. *International Journal of Man–Machine Studies 32,* 385–398.

Carroll, J. M., & Rosson, M. B. (1987). Paradox of the active user. In J. M. Carroll (Ed.), *Interfacing thought: Cognitive aspects of human–computer interaction* (pp. 80–111). Bradford/MIT Press.

Charney, D., Reder, L., & Kusbit, G. (1990). Goal setting and procedure selection in acquiring computer skills: A comparison of problem solving and learner exploration. *Cognition and Instruction, 7,* 323–342.

Draper, S. W. (1986). Display managers as the basis for user–machine communication. In D. A. Norman & S. W. Draper (Eds.), *User centered system design* (pp. 339–352). Lawrence Erlbaum Associates, Inc.

Franzke, M. (1995). Turning research into practice: Characteristics of display-based interaction. *Proceedings of the CHI'95 Conference on Human Factors in Computing Systems,* 421–427. New York: ACM.

Furnas, G. W., Landauer, T. K., Gomez, L. W., & Dumais, S. T. (1987). The vocabulary problem in human–system communication. *Communications of the ACM, 30,* 964–971.

Howes, A. (1994). A model of the acquisition of menu knowledge by exploration. *Proceedings of the CHI'94 Conference on Human Factors in Computing Systems,* 445–451. Boston: ACM.

Howes, A., & Payne, S. J. (1990). Display-based competence: Towards user models for menu-driven interfaces. *International Journal of Man–Machine Studies, 33,* 637–655.

Howes, A., & Young, R. M. (1996). Learning consistent, interactive and meaning-ful device methods: A computational model. *Cognitive Science, 20,* 301–356.

John, B. E., Vera, A. H., & Newell, A. (1994). Toward real-time GOMS: A model of expert behavior in a highly interactive task. *Behavior and Information Technology, 13*(4), 255–267.

Kellogg, W. A. (1987). Conceptual consistency in the user interface: Effects on user performance. *Proceedings of the INTERACT'87 Conference on Human–Computer Interaction,* 389–394. Amsterdam, The Netherlands: Elsevier.

Kintsch, W. (1970). Models for free recall and recognition. In D. A. Norman (Ed.), *Models of human memory* (pp. 333–374). New York: Academic.

Kintsch, W. (1988). The role of knowledge in discourse comprehension: A con-struction–integration model. *Psychological Review, 95,* 163–182.

Kirschenbaum, S. S., Gray, W. D., & Young, R. M. (1996). Cognitive architectures and HCI [Workshop report]. *SIGCHI Bulletin, 28*(2), 18–21.

Kitajima, M. (1989). A formal representation system for the human–computer interaction process. *International Journal of Man–Machine Studies, 30,* 669–696.

Kitajima, M., & Polson, P. G. (1995). A comprehension-based model of correct performance and errors in skilled, display-based, human–computer interaction. *International Journal of Human–Computer Studies, 43,* 65–99.

Laird, J. E. (1986). Universal subgoaling. In J. E. Laird, P. S. Rosenbloom, & A. Newell (Eds.), *Universal subgoaling and chunking: The automatic generation and learning of goal hierarchies* (pp. 1–131). Boston: Kluwer.

Lansdale, M. W. (1991). Remembering about documents: Memory for appearance, format, and location. *Ergonomics, 34,* 1161–1178.

Lansdale, M. W. (1995). *Modelling errors in the recall of spatial location* (Technical Report). Loughborough, England: Loughborough University of Technology, Cognitive Ergonomics Research Group.

Larkin, J. H. (1989). Display-based problem solving. In D. Klahr & K. Kotovsky (Eds.), *Complex information processing: The impact of Herbert A. Simon* (pp. 319–342). Hillsdale, NJ: Lawrence Erlbaum Associates, Inc.

Larkin, J. H., & Simon, H. A. (1987). Why a diagram is (sometimes) worth ten thousand words. *Cognitive Science, 11,* 65–99.

Lee, A. Y., Foltz, P. W., & Polson, P. G. (1994). Memory for task-action map-pings—Mnemonics, regularity and consistency. *International Journal of Hu-man–Computer Studies, 40,* 771–794.

Lewis, C. H. (1988). Why and how to learn why: Analysis-based generalization of procedures. *Cognitive Science, 12,* 211–256.

May, J., Barnard, P. J., & Blandford, A. E. (1993). Using structural descriptions of interfaces to automate the modelling of user cognition. *User Modelling and User-Adapted Interaction, 3,* 27–64.

Mayes, J. T., Draper, S. W., McGregor, M. A., & Oatley, K. (1988). Information flow in a user interface: The effect of experience and context on the recall of MacWrite screens. In D. M. Jones & R. Winder (Eds.), *People and Computers IV* (pp. 275–289). Cambridge, England: Cambridge University Press.

Miller, C. S., & Laird, J. E. (1992). *A simple symbolic algorithm for incremental concept acquisition.* Ann Arbor: University of Michigan, Artificial Intelligence Laboratory.

Miller, C. S., Lehman, J. F., & Koedinger, K. R. (1997). *Goal-directed learning in microworld interaction.* Manuscript submitted for publication.

:, G. A., & Charles, W. G. (1991). Contextual correlates of semantic similarity. *Language and Cognitive Processes, 6*(1) 1–28.

Mitchell, T. M. (1986). Explanation-based generalization: A unifying view. *Machine Learning 1,* 47–80.

Moore, J., & Newell, A. (1974). How car Merlin understand? In L. W. Gregg (Ed.), *Knowledge and cognition* (pp. 201–252) Potomac, MD: Lawrence Erlbaum Associates, Inc.

Naveh-Benjamin, M. (1987). Coding of spatial information—An automatic process. *Journal of Experimental Psychology: Learning, Memory, and Cognition, 13,* 595–605.

Newell, A. (1969). Heuristic programming: Ill-structured problems. In J. Aronofsky (Ed.), *Progress in Operations Research III* (pp. 363–414). New York: Wiley.

Newell, A. (1990). *Unified theories of cognition.* Cambridge, MA: Harvard University Press.

Payne, S. J. (1985). *Task-action grammars: The mental representation of task languages in human–computer interaction.* Unpublished doctoral dissertation, University of Sheffield, Sheffield, England.

Payne, S. J. (1991). Display-based actior at the user interface. *International Journal of Man–Machine Studies, 35,* 275–289.

Payne, S. J., & Green, T. R. G. (1986) Task-action grammars: A model of the mental representation of task languages. *Human–Computer Interaction, 2,* 93–133.

Payne, S. J., & Green, T. R. G. (1989). The structure of command languages: An experiment on task-action grammar. *International Journal of Man–Machine Studies, 30,* 213–234.

Polson, P. G., & Lewis, C. H. (1990). Theory-based design for easily learned interfaces. *Human–Computer Interaction, 5,* 191–220.

Richman, H. B., Staszewski, J. J., & Simon, H. A. (1995). Simulation of expert memory using EPAM IV. *Psychological Review, 2,* 305–330.

Rieman, J. F. (1994). *Learning strategies and exploratory behavior of interactive computer users* (CU–CS–723–94). Unpublished PhD thesis, University of Colorado, Boulder.

Robert, J. M. (1987). Learning a computer system by unassisted exploration. *Proceedings of the INTERACT'87 Conference on Human–Computer Interaction,* 651–656. Amsterdam: Elsevier.

Rosenbloom, P. S., & Aasman, J. (1990). Knowledge level and inductive uses of chunking (EBL). *Proceedings of the National Conference on Artificial Intelligence,* 821–827. Boston: AAAI.

Rosenbloom, P. S., & Laird, J. E. (1986). Mapping explanation-based generalization onto Soar. *Proceedings of the AAAI–86 Conference on Artificial Intelligence,* 561–567. Los Altos, CA: Kaufmann.

Rosenbloom, P. S., Laird, J. E., & Newel, A. (1987). Knowledge level learning in Soar. *Proceedings of the AAAI–87 Conference on Artificial Intelligence,* 499–504. Los Altos, CA: Kaufmann.

Rosenbloom, P. S., Laird, J. E., & Newell, A. (1988). The chunking of skill and knowledge. In B. A. G. Elsendoorn & H. Bouma (Eds.), *Working models of human perception* (pp. 391–410). London: Academic.

Vera, A. H., Lewis, R. L., & Lerch, F. J. (1993). Situated decision-making and recognition-based learning: Applying symbolic theories to interactive tasks.

Proceedings of the Conference of the Cognitive Science Society, 84–95. Hillsdale, NJ: Lawrence Erlbaum Associates, Inc.

Young, R. M. (1982). Architecture-directed processing. *Proceedings of the Conference of the Cognitive Science Society,* 164–166. Hillsdale, NJ: Lawrence Erlbaum Associates, Inc.

HUMAN–COMPUTER INTERACTION, 1997, Volume 12, pp. 345–389

A Comprehension-Based Model of Exploration

Muneo Kitajima
National Institute of Bioscience and Human-Technology, Japan

Peter G. Polson
University of Colorado

ABSTRACT

The linked model of comprehension-based action planning and instruction taking (LICAI) simulates performing by exploration tasks using applications hosted on systems with graphical user interfaces. The tasks are given to the user as written exercises containing no information about the correct action sequences. LICAI's comprehension and action-planning processes are based on Kintsch's construction–integration (C–I) theory for text comprehension. The model assumes that comprehending instructions is a strategic process; instruction texts must be elaborated using specialized strategies that guide goal generation. LICAI comprehends the instructions and generates goals that are then stored in memory. The action-planning processes are controlled by goals retrieved from memory. Representations of goals that can guide exploration are restricted by the C–I architecture. The model predicts that successful exploration requires linking of the goal representation with the label on the

Muneo Kitajima is a cognitive scientist with an interest in cognitive modeling of various aspects of human–computer interaction; he is a Senior Researcher in the Department of Human Informatics at the National Institute of Bioscience and Human-Technology, Agency of Industrial Science and Technology, Ministry of International Trade and Industry, Japan. **Peter G. Polson** is a cognitive psychologist with interests in computer simulation models of skill acquisition and usability evaluation methods; he is Professor of Psychology and a Fellow of the Institute of Cognitive Science at the University of Colorado at Boulder.

CONTENTS

correct object. The model is evaluated by comparing its predictions with results from an experimental study of learning by exploration by Franzke (1994, 1995). We discuss the implications of LICAI for designing instruction materials and interfaces that facilitate exploration.

1. INTRODUCTION

This article presents a model, the linked model of comprehension-based action planning and instruction taking (LICAI),[1] of the cognitive processes involved in following incomplete instructions for using applications hosted on systems with graphical user interfaces. The instructions describe a task to be performed but do not contain any information about action sequences. Thus, a new user must perform the task by exploration. LICAI simulates the processes involved in comprehending the instructions and then generates, by exploration, an action sequence that performs the task.

A focus of recent research on skill acquisition in human–computer interaction (HCI) has been on learning by exploration (Carroll, 1990; Howes, 1994; Rieman, 1994, 1996; Rieman, Young, & Howes, 1996). Experienced users of a given environment (e.g., Macintosh or Windows 95) learn new applications or extend their knowledge of applications they already know by task-oriented exploration. Formal training is rarely available, and there are usability problems with training and reference documentation (see Compeau, Olfman, Sein, & Webster, 1995). Most users prefer to acquire new skills by exploration as they perform new tasks relevant to their work (Carroll, 1990; Rieman, 1994, 1996).

Most current models of exploration are based on problem-solving architectures like Soar (Newell, 1990; Rieman et al., 1996) or ACT–R (Anderson, 1993; Rieman, 1994; Rieman, Lewis, Young, & Polson, 1994). Many of these models attempt to account for the label-following strategy (Engelbeck, 1986; Polson & Lewis, 1990; Franzke, 1994, 1995), which is one of the most frequently used problem-solving heuristics by users at all levels of expertise. *Label following* involves using the overlap between users' goals and labels on menus, buttons, and other interface objects to guide search during exploration. Interface objects with overlapping labels are acted on (e.g., by dropping a menu) in attempts to discover correct action sequences.

LICAI, in contrast, focuses on goal formation. Previous models of the label-following strategy have taken goals as given and then described the

---

1. When LICAI is pronounced *lick eye* or *lee chi* (χ), the pronunciation represents the two-kanji-character Japanese word 理解 ("comprehension").

resulting search behavior (e.g., Rieman et al., 1996). However, models of exploration should go further and define the processes that generate the goals that guide exploratory behavior. Usually, tasks are given to users through various forms—narratives, written instruction texts, help messages, graphic images, and combinations of these forms. Whatever form is used to specify the task, the user must comprehend the meaning of an initial task description and then formulate goals that guide interaction with the interface. In LICAI, comprehension (i.e., goal formation) and exploration processes are modeled using the construction–integration (C–I) cognitive architecture (Kintsch, in press). Goal formation is simulated using a version of Kintsch's (1988) model of word-problem solving. Exploration is simulated using an extension of Kitajima and Polson's (1995) display-based action-planning model.

The C–I architecture has evolved from a detail model of skilled, text comprehension—a highly automated collection of cognitive processes that make use of massive amounts of knowledge stored in long-term memory (Kintsch, 1988, in press). This is a very different foundation from the other cognitive architectures that underlie other models described in this special issue of *Human–Computer Interaction*. For example, one of the primary foundations of Soar (Newell, 1990) was the general problem solver (Ernst & Newell, 1969), a model of deliberate cognition (i.e., problem solving and action planning) in situations in which the problem solver has limited background knowledge.

1.1. Outline of LICAI (Linked Model of Comprehension-Based Action Planning and Instruction Taking)

What motivated us to develop LICAI—a comprehension-based model of exploration—was Franzke's (1994, 1995) experiments in which experienced Macintosh users were introduced to a new application. Participants had to learn the application by carrying out a series of written exercises by exploration. The instructions carefully described a task but did not provide information about an action sequence that would accomplish the task. LICAI simulates such users.

LICAI is a synthesis and extension of previous models that have been developed based on the C–I architecture (Kintsch, in press). The processes of comprehending task instructions to form goals are modeled using an extended version of Kintsch's (1988, in press [chap. 10]) model of word-problem solving. LICAI incorporates a goal-selection process to deal with situations in which the instructions describe several subtasks, and the reader must retrieve from long-term memory the goal for each task cued by information generated by the application interface. The retrieval process is modeled by a version of the Kintsch and Welsch (1991) model of cued recall. The retrieved goal controls the exploration processes that are modeled by an extended version of Kitajima and Polson's (1995) model of skilled, display-based HCI.

The C–I theory, however, is an incomplete cognitive architecture (cf. Soar [see Newell, 1990] or ACT–R [see Anderson, 1993]), because it does not incorporate explicit mechanisms for learning and problem solving. Wharton and Kintsch (1991) and Kintsch (in press, chap. 11) outlined how the theory could be extended. The action-planning models (Kitajima & Polson, 1995; Mannes & Kintsch, 1991) contain the elements of a problem-solving architecture, including goal formation, action selection, and the ability to react to the consequences of actions.

Although LICAI does not incorporate search mechanisms, our results do refine and extend Franzke's (1995) analysis based on the model of exploration (Polson & Lewis, 1990) underlying the "cognitive walk-through" (Polson, Lewis, Rieman, & Wharton, 1992; Wharton, Rieman, Lewis, & Polson, 1994). Our simulation accounts for the initial success or failure of the goal-formation process, and we show that there are strong constraints on the exact goal form that enables the action-planning processes to generate the correct action sequence for an exercise. Thus, LICAI partitions instructions given to Franzke's participants into (a) those that enable the model to do the task with little or no trial-and-error search and (b) those that the model cannot perform because it cannot formulate effective goals from the instructions.

1.2. The Graph Task and Franzke's Experimental Procedure

Franzke (1994) and the simulation reported in this article, as well as Kitajima and Polson (1995) and Rieman et al. (1996), used the following task. Participants graph a set of data contained in an application document as a line graph. The graphing task is divided into two subtasks, and Franzke used different instructions for each subtask. The first subtask is to create a default line graph. The instructions provide participants with the information necessary to perform the several steps necessary to generate the default graph, including variable names and graph style. Franzke's participants, however, were faced with two problems—extracting the information that would enable them to discover these steps and sequencing this information so that the steps would be attempted in the correct order. The second subtask is to edit the default line graph created during the first subtask. The edits are to be done in a specific order, and the instructions are short and telegraphic. The third problem for participants is to make the inferences necessary to generate a comprehensible task description that could successfully guide exploration. The goal-formation and goal-selection processes incorporated into LICAI solve these three problems posed by Franzke's instructions.

Figure 1. Variable-selection dialog box from Cricket Graph I.

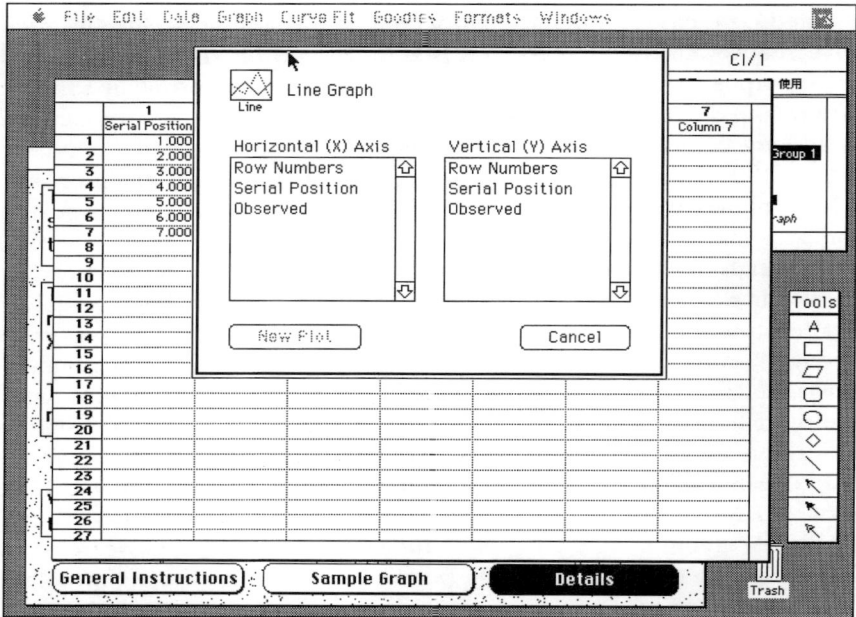

Franzke's participants were experienced Macintosh users who had never used a graphing application. They performed two isomorphic tasks in which different sets of variables, file names, and graph styles were provided, and each task was accomplished by using one graphing application—Cricket Graph I[2] or III[3] or one of two forms of Excel 3.0[4]—by exploration. If a participant had not progressed after 2 min on a particular step, he or she was given brief hints like "Select line graph from the graph menu" and "Double-click on legend text."

One group of Franzke's participants were asked to accomplish the task using Cricket Graph I. The data are stored in a document file containing a spreadsheet with columns labeled with variable names. The task is to graph the data in the column labeled *Observed* as a function of numbers in the column labeled *Serial Position.* The initial step is to double-click on the document icon to launch the application and display the data. The next step is to pull down the *Graph* menu and release on the item *Line* so that the dialog box with two scrolling lists appears (see Figure 1). The column labels are displayed in two scrolling lists, representing the x- and y-axes.

2. CA Cricket Graph, Version 1.3.2, 1989.
3. CA Cricket Graph III, Version 1.01, 1992.
4. Microsoft Excel, Version 3.0, 1990.

Serial Position and *Observed* appear in both lists. The dialog box partially occludes the table, but the column of numbers labeled *Serial Position* is visible in the background. To plot *Observed* as a function of *Serial Position,* participants must click on and highlight *Serial Position* in the x-axis scrolling list and *Observed* in the y-axis scrolling list. The final step is to click on *New Plot,* causing a default version of the graph to be displayed. The remaining steps in the task involve editing this default version.

1.3. Outline of the Article

In Section 2, we describe Kintsch's (1988, in press) C–I architecture, its applications to several cognitive processes, and their relations to LICAI. We introduce a single C–I cycle and then describe several processes defined by multiple C–I cycles. These processes include text comprehension (Kintsch, 1988), word-problem solving (Kintsch, 1988), memory retrieval (Kintsch & Welsch, 1991), action planning (Mannes & Kintsch, 1991), and skilled, display-based HCI (Kitajima & Polson, 1995). Section 3 describes LICAI by tracing a simulation of Franzke's (1994, 1995) task. In Section 4, we evaluate LICAI using Franzke's (1995) results; in Section 5, we review recent models of exploration in HCI in light of our results. In Section 6, we summarize the implications of our results for practice, especially those for the design of interfaces and training materials that support exploration (Carroll, 1990; Wharton et al., 1994).

2. THE CONSTRUCTION–INTEGRATION (C–I) ARCHITECTURE

The cognitive processes specified in LICAI are implemented using the C–I architecture (Kintsch, 1988, in press; Wharton & Kintsch, 1991). As mentioned earlier, LICAI is a synthesis and extension of several models that have evolved from Kintsch's (1988) original model of text comprehension. This section presents a general introduction to the C–I architecture and the various models realized using the architecture. Section 2.1 provides an overview. Section 2.2 describes basic ideas underlying the C–I architecture. Section 2.3 focuses on the cognitive processes modeled on the C–I architecture and describes how these processes are incorporated into LICAI.

Our treatment is very different from Kintsch's (in press). We characterize the basic C–I cycle as roughly analogous to the decision cycle of Soar (Newell, 1990), and we regard the various models as special cases of this general process. Soar views all cognitive behavior as problem solving (search through a problem space). However, the models described in this section actually evolved from a model of skilled reading. The title of Kintsch's (in press) book is *Comprehension: A Paradigm for Cognition.* Kintsch

views LICAI and the Kitajima and Polson (1995) model of skilled performance as generalizations of a theory of comprehension.

2.1. Overview

The basic process assumed by the architecture is a two-phase C–I cycle.
The cycle is a cognitive process that manipulates knowledge represented
as propositions. Three ideas characterize the C–I cycle.

First, a cycle selects between alternatives. Thus, the C–I cycle is roughly
analogous to the decision cycle of Soar (Newell, 1990), which selects
among alternative operators. However, Soar uses different processes and
representations.

Second, there are different C–I cycles defined by various models (see
subsections of Section 2.3):

- Text-comprehension cycles select between alternative interpreta-
  tions of sentences (Kintsch, 1988).
- Problem-model construction cycles, a strategic form of text-compre-
  hension cycles, generate representations that control solution of a
  problem described in a text (Kintsch, 1988).
- Memory-retrieval cycles select one out of many possible representations from long-term memory in a cued-recall paradigm
  (Kintsch & Welsch, 1991).
- Action-selection cycles select between alternative actions (Mannes &
  Kintsch, 1991).
- Attention cycles select a subset of the information in a complex
  visual display for further processing by action-selection cycles (Kita-
  jima & Polson, 1995).

Third, many processes involve multiple C–I cycles. These processes
may include the same cycle or a series of different cycles. For example,
reading a short paragraph is modeled as a series of text-comprehension
cycles (Kintsch, 1988). Reading a text describing a word problem and
building a representation that controls successful problem solving is simulated as a series of problem-model construction cycles (Kintsch, 1988).
Skilled performance of a task using a computer with a graphical user
interface is simulated as a series of pairs of attention and action-selection
cycles (Kitajima & Polson, 1995).

LICAI simulates the processes involved in reading instructions and
performing a task using three types of cycles—problem-model construction cycles for generating goals, memory-retrieval cycles for selecting a
goal, and one or more pairs of attention–action-selection cycles for finding
an action sequence that satisfies the selected goal.

2.2. A Single C–I Cycle

A C–I cycle consists of a construction phase and an integration phase. The construction phase generates a network of propositions that contains the surface representation of a stimulus (text or visual display), alternative interpretations of the surface representation, possible actions, or other alternatives. The network also incorporates the knowledge necessary for selecting among the alternatives. This knowledge includes goals, information retrieved from long-term memory, and information carried over from previous C–I cycles.

Various models within the C–I architecture assume parsers that map surface representations into propositional semantic representations (text in Kintsch, 1988; visual displays in Kitajima & Polson, 1995). Generating alternatives is assumed to be a bottom-up, weakly constrained, rule-based process. The rules are not context sensitive; thus, the alternatives represented in the network may not be consistent with the current context.

The alternatives and all other information in the network are represented as propositions (Kintsch, 1974, 1988, in press; van Dijk & Kintsch, 1983). A proposition is a unit of knowledge that is represented as an ordered tuple. The first element is a predicate, and the remaining elements are arguments of the predicate. For example, the goal "Graph data" is represented by the proposition (PERFORM GRAPH DATA). The predicate PERFORM defines the proposition as representing a goal of the form, "Perform an action on an object," in which the first argument is the action and the second is the object. Propositions in the network are linked by their shared arguments. Thus, any proposition containing the arguments GRAPH or DATA will be linked to (PERFORM GRAPH DATA).

The integration phase selects an alternative by integrating information represented in the network generated during the construction phase. Integration can be thought of as a constraint-satisfaction process. The network of interconnected propositions defines a collection of constraints that are satisfied by the selected alternative. Integration is performed using a spreading activation process. In this process, the nodes in the network can be partitioned into sources of activation, targets of activation, and links between sources and targets. Goals and the representation of the current context (i.e., text or visual display) are typical sources, and the targets are alternatives. The linking information comes from long-term memory and other sources, and the spreading activation process is controlled by the pattern of links in the network. When the integration phase terminates, the most highly activated alternative represents the result of the C–I cycle that satisfies the constraints. Because propositions in the network are linked by shared arguments, the pattern of argument overlap plays a key role in the results of the integration phase.

2.3. Specific Versions for Reading, Memory Retrieval, and Action Planning

This section summarizes various models that are based on the C–I architecture and that we modified and incorporated into LICAI. Our descriptions are based on abstract characterization of the C–I cycle presented in Section 2.2.

Text-Comprehension Cycles

The initial version of the architecture was a C–I model of text comprehension (Kintsch, 1988), and this model retained many basic assumptions from Kintsch's earlier work on text comprehension (Kintsch, 1974; Kintsch & van Dijk, 1978; van Dijk & Kintsch, 1983). Reading a text is simulated by a series of C–I cycles. During each construction phase, the model takes as input a representation of key elements of the text comprehended so far and a propositional representation of the next sentence or major sentence fragment. The model incorporates knowledge relevant to the input text as elaborations, which come from two sources: (a) propositions retrieved from long-term memory by a stochastic, associative retrieval process (Raaijmakers & Shiffrin, 1981) and (b) inferences generated by schemata triggered by propositions in the original input text. At the end of the construction phase, the network incorporates multiple interpretations of the input text.

During the integration phase, the model selects an interpretation of the input sentence that satisfies the constraints defined by the network. The sources of activation are the carried-over propositions and the propositions representing the input text. The targets are all the propositions in the network. The most highly activated subset of propositions in the network represent the reader's interpretation of the text. These nodes are typically the most highly interconnected subset.

Problem-Model Construction Cycles

One of the fundamental issues for a theory of comprehension is to show how readers can use information contained in texts to solve problems or how users can perform tasks described in instructions. The text must be transformed into a representation that can mediate successful problem solving or action planning. Kintsch and his collaborators (e.g., Kintsch & Greeno, 1985) have developed models of word-problem solving that transform texts describing arithmetic or algebra problems into special forms that specify arithmetic or algebra operators. Typically, the reader must elaborate the original text with specialized inferences that guide problem solving. LICAI incorporates analogous comprehension processes that generate goals that control action-planning processes. Section 3.3 describes the processes in detail.

Kintsch (1988) showed that the C–I architecture can be extended to problem-model construction for arithmetic word problem solving by incorporating versions of assumptions made by Kintsch and Greeno (1985) and others (Cummins, Kintsch, Reusser, & Weimer, 1988; Delarosa, 1986; Fletcher, 1985). Kintsch's (1988) arithmetic word model and the earlier work all incorporate the assumption that reading is a strategic process (van Dijk & Kintsch, 1983). Strategies generate inferences required to construct a representation that satisfies a reader's goals (e.g., solve a word problem). The knowledge used by the strategies is represented as *comprehension schemata*.

Kintsch (1988) incorporated the comprehension schemata needed to solve word problems by assuming that they operate during the construction phase by adding propositions to the network. These schemata elaborate the original text propositions by generating special interpretations of the text as well as representations that mediate problem solving. For example, one schema identifies noun phrases as sets, and comprehension schemata generate problem models by specifying alternative hypotheses about the role of each set in a problem description (e.g., part set, whole set). Representations of all possible problem models are incorporated into the network. The alternative problem models are linked to other components of the network, including information about time, order of possession, location, and other aspects of the situation described in the problem text.

At the end of the integration phase, the most highly activated problem model is the one consistent with the situation described in the problem text. In the spreading activation process, the original text propositions are the sources of activation, and the alternative problem models are the targets. Propositions describing the situation link the original problem text with problem models if they are consistent. These supporting links define constraints that lead to the selection of one problem model. The selected problem model is then operated on by problem-solving mechanisms (arithmetic, algebra, or action planning) to generate a solution to the problem described in the text.

Memory-Retrieval Cycles

Franzke's (1994) instructions for the first subtask, creating a default graph, contained information used in several different steps required to perform this subtask. The comprehension processes incorporated into LICAI read the complete text before attempting to perform the subtask and store goals and other information needed to perform the subtask in long-term memory. LICAI retrieves and sequences these goals using the displays generated by the application as the retrieval cues.

Kintsch and Welsch (1991) showed that the C–I architecture can be extended to account for memory for text. The model assumes that each text-comprehension cycle is followed by a process that transfers the results of the comprehension cycle to episodic memory. Retrieval from episodic memory is modeled as a C–I cycle. During the construction phase, a retrieval cue is linked to the episodic memory network. During the integration phase, the retrieval cue (activation source) activates the nodes in the episodic memory network (targets). Kintsch and Welsch (1991) showed that the activation value of each node in the episodic memory correlates with the speed of recall of that node. LICAI uses a modified version of the Kintsch and Welsch (1991) model to simulate retrieval processes, described in Section 3.5.

Action-Selection Cycles

Mannes and Kintsch (1991) extended the C–I model of text comprehension to action planning using HCI as a task domain by assuming that text comprehension and action planning are similar tasks. Readers integrate information from diverse sources to select one out of many alternative interpretations of a text. Similarly, users, as action planners, must integrate their goals and information from diverse sources and select one out of many competing actions.

Mannes and Kintsch's (1991) action-selection cycle modified the basic C–I architecture by adding rule-like action representations called *plan elements*. The condition of a plan element is a set of propositions that describe its preconditions. The action of a plan element is one or more propositions that describe the consequences of performing a plan element in a simulated task. These plan elements are incorporated into the network as nodes represented by object–action pairs (e.g., EDIT FILE ⌃ LETTER, a proposition that represents the action EDIT applied to a FILE named LETTER).

During the construction phase, a network is generated to include propositions that represent the task description (i.e., the user's goal), the task context (including the consequences of previous cycles), knowledge retrieved from long-term memory, and plan elements. These plan elements are generated by the following procedure: For each cycle, the model is given a list of three objects in the current task context, and each of the three objects is combined with all possible actions. The propositions in the network are then connected by shared arguments.

Mannes and Kintsch (1991) also made important modifications to the integration phase. During the activation process, the model spreads activation over the network using the same activation process described by Kintsch (1988); the activation sources are the propositions describing the task and the current task context, and the targets are the plan elements. A decision process operates after the spreading activation process. The model evaluates the conditions of highly activated plan elements in order of their activation values. A condition is true if all of its propositions are found in the network. The model executes the most highly activated plan element whose condition is true.

Attention Cycles Combined With Action Cycles

Mannes and Kintsch (1991) simulated a task environment in which there were few objects in the world that were candidates for possible actions. Kitajima and Polson (1995) extended Mannes and Kintsch's (1991) model of action planning to display-based HCI. Kitajima and Polson incorporated into their model a complex representation of a large-format display containing tens of objects that are targets for possible actions.

Kitajima and Polson (1995) simulated skilled users who possess the knowledge necessary to correctly perform a task. They made two major additions to Mannes and Kintsch's (1991) action-planning model. First, they assumed that skilled users have two level representations of familiar tasks. Second, they added an attention process that focuses the action-planning processes' attention on just three of the possible targets for action. Their model was mapped onto Hutchins, Hollan, and Norman's (1986) analysis of direct manipulation, which provides a framework to describe action planning as a goal-driven process. The action-planning process evaluates the consequences of the last action and then generates the next action to be executed (see Figure 2). Kitajima and Polson simulated the process of a user's comprehending the display (Hutchins et al.'s, 1986, evaluation stage) by using the elaboration process of the construction phase.

In the next three subsections, we describe Kitajima and Polson's (1995) model of skilled HCI in more detail and show how LICAI extends it to simulate an experienced user's attempting to generate correct action sequences by exploration.

Task and Device Goals.   Kitajima and Polson's (1995) action-planning model assumes that skilled users have a schematic representation of the task in the form of a hierarchical structure involving *task goals* and *device goals* (Payne, Squibb, & Howes, 1990). They assumed that each task goal is associated with one or more device goals. A task goal is of the form, "Perform a task action (e.g., graph, delete, put) on a task object (e.g., data, word, variable)." One or more device goals are associated with each task goal and specify states of specific screen objects that must be achieved to satisfy an associated task goal. Examples include specifying that a variable name in a scrolling list, a menu item, or the like should be highlighted. Device goals control the action-planning processes.

Figure 2. The action-planning model of correct performance and errors in skilled, display-based HCI (Kitajima & Polson, 1995) mapped onto Norman's (1986) action theory framework.

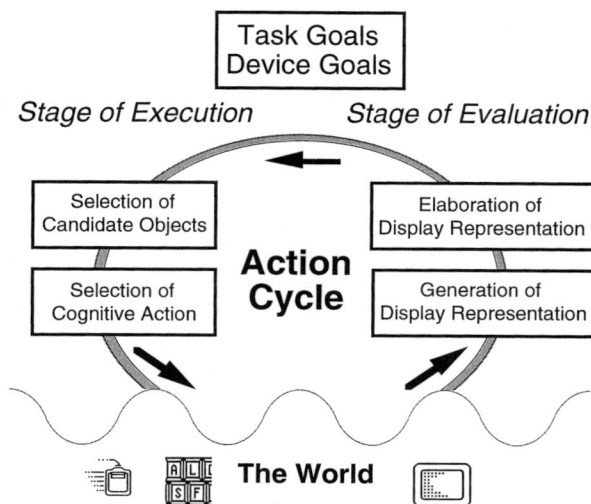

However, LICAI simulates a new user of an application. LICAI assumes that the instruction-comprehension processes generate task goals. The goal-selection process retrieves a task goal from long-term memory and passes it to the action-planning processes. The action-planning processes must attempt to generate an action sequence that will accomplish the task goal without knowledge of the device goals. We assume that device goals are learned by interacting with the application.

The Evaluation Stage. Kitajima and Polson's (1995) action-planning model is given a representation of a new display in the form of a large collection of screen objects; each screen object is described by several propositions. These descriptions provide limited information about the identity of each object and its appearance, including visual attributes (e.g., color, highlighting). The model simulates Hutchins et al.'s (1986) evaluation stage (shown in Figure 2) by elaborating the display representation with knowledge retrieved from long-term memory. The retrieval cues are the task and device goals and the propositions representing the current display. The probability that a cue retrieves a particular proposition representing a piece of knowledge from long-term memory is computed by Raaijmakers and Shiffrin's (1981) model.

The propositions in long-term memory represent knowledge about the screen objects. For example, if Object23 is the scrolling list item labeled *Serial Position,* then the following knowledge items are stored in long-term memory about Object23: Object23 has-label Serial_Position; Object23 is-part-of Line-Graph-Dialog-Box; Object23 can-be-pointed-at; and Object23 can-be-selected. The elaboration process is stochastic, and Kitajima and Polson (1995) discussed in detail the predictions and implications that follow from this stochastic elaboration process.

The Execution Stage.   Kitajima and Polson (1995) model the execution stage of Hutchins et al.'s (1986) framework as a pair of an attention cycle and an action-selection cycle.

Attention Cycles.   The attention cycle selects three screen objects as possible candidates for action to be carried over to the succeeding action-selection cycle. During the construction phase, a network is generated consisting of nodes representing all screen objects, the task and device goals, the elaborated display from the evaluation stage, and candidate object nodes of the form, Screen-Object-X is-attended. During the integration phase, the sources of activation are the task and device goals and the screen objects, and the targets are the candidate object nodes. When the spreading activation process terminates, the model selects the three most highly activated candidate object nodes. These nodes represent screen objects to be attended to during the action-selection cycle.

The activation pattern is determined largely by two factors: (a) the links from the task and device goals to propositions in the network that share arguments with the goals and (b) the number of propositions necessary to link goals to candidate objects. As a result, the attention cycle selects candidate objects closely related to the task and device goals.

Device goals can directly specify a screen object and thus can be linked directly to the screen object represented in the network. Task goals can be linked indirectly to screen objects through labels. Continuing with the previous example, consider the following task goal and its associated device goal:

Task goal: perform 'plot Serial_Position on x-axis'
Device goal: realize Object23 is-highlighted

The correct candidate object node Object23 is-attended has a direct link with the device goal. On the other hand, the task goal can be linked to the correct candidate object node via a proposition in the form Object23 has-label Serial_Position. Note that the device goal focuses the model's attention on the correct screen object. The task goal may or may not perform this function. The only way it can is for an element of the task goal representation to match the label on the correct screen object.

Action-Selection Cycles.   The action-selection cycle works like Mannes
and Kintsch's (1991); however, the actions are not commands but operations on screen objects. The action-selection cycle selects an action to be
performed on one of the three candidate objects. During the construction
phase, the model generates a network that includes the task and device
goals, all screen objects, the elaborated display, and representations of all
possible actions on each candidate object (i.e., plan elements). Examples
would include single-click Object23, move Object23, and the like.

During the integration phase, the sources are the task and device goals
and the screen objects, and the targets are the nodes representing plan
elements. At the end of the integration phase, the model selects the most
highly activated plan element whose preconditions are satisfied as the next
action to be executed. The process is dominated by the same two factors
as in the attention cycle. However, the relevant interaction knowledge
must be retrieved during the evaluation stage satisfying preconditions for
the correct action. For example, the model must retrieve from long-term
memory the fact that objects in a scrolling list can be selected (i.e., Object23
can-be-selected).

LICAI's action-planning processes operate without device goals to
simulate a user who is dealing with a novel application. However, if LICAI
manages to focus on the correct screen object, it is likely to perform the
correct action. Recall that LICAI simulates skilled Macintosh users. The
model's knowledge of the Macintosh interface conventions enables it to
identify the attended-to screen object and retrieve from long-term memory
knowledge about correct action for this object.

3. LICAI

This section provides a detailed description of LICAI. The model
integrates extended versions of problem-model construction, retrieval,
attention, and action-selection cycles to simulate goal formation, goal
selection, and exploration. Task goals coordinate the operations of these
cycles and control the three processes defined in LICAI. Section 3.1
describes requirements for the representation of task goals that can lead
the action-planning processes to successful exploration. Then, we explain
LICAI by tracing a simulation of the task to create a default version of a
line graph. Section 3.2 introduces the task and the instructions. Section 3.3
presents the specialized comprehension schemata used for elaborating the
text representations with propositions describing task goals. Section 3.4
describes the simulation of the goal-formation processes and shows how
the instructions are processed by applying the comprehension schemata
and how the results of comprehension are stored in memory. Section 3.5

describes the simulation of the exploration processes, including retrieval of the task goal and action planning.

3.1. Exploration Guided by Task Goals

LICAI simulates a skilled user of the Macintosh interface learning a novel application. The goal-formation processes of LICAI generate possible task goals from the instructions using the comprehension schemata. For experienced users, device goals control the generation of action sequences that satisfy their associated task goals. However, new users do not have device goals, and successful exploration in LICAI is controlled solely by a task goal. It is assumed that device goals, which are descriptions of the states of specific screen objects, must be learned by successful interaction with the application.

Kitajima (1996) reported on a series of simulation experiments that determined the constraints on the exact representation of task goals that mediate successful action planning. First, Kitajima found that the action-planning model always generated correct actions when given correct device goals. This was true even when there was no task goal. A device goal specifies a screen object and a desired attribute of the object. For example, the variable name *Serial Position* in the x-axis scrolling list must be highlighted to accomplish the task goal of perform 'plot Serial_Position on x-axis'. The corresponding device goal is realize Object23 is-highlighted, where Object23 is the model's representation of column label *Serial Position* in the x-axis scrolling list. The direct link between the device goal and the correct screen object caused the attention cycle to include the correct screen object in the list of the three candidate screen objects for the next action. When selecting an object–action pair during the action-selection cycle, the model usually chose the correct action on the most highly activated screen object. The specification of the desired attribute guides selection of the correct action because the model has knowledge of the consequences of an action on an object.

Second, Kitajima found that the model could generate the correct actions when given a task goal without device goals. However, the action-planning process was not as robust as when device goals are provided. In order for the model to make correct object–action selections, the task goal had to be specific enough to establish a link with the correct screen object. For example, the representation of the task goal perform 'plot Serial_Position on x-axis' has a link to Object23 through its label. This link enables the action-planning process to select the correct screen object during the attention cycle. However, an equivalent task goal, perform 'plot Observed as_a_function_of Serial_Position', will not work because the screen label x-axis was missing. Because there are three screen objects in Figure 1 with the label *Serial Position,* the link to the label of the x-axis scrolling list is necessary for the

Figure 3. **Instruction texts: A version from Franzke (1994).**

Instruction Texts	Cycle Number
In this experiment, you are going to learn a new Macintosh application, Cricket Graph, by exploration.	1
The task you are going to perform will be presented to you as a series of exercises.	2
The data you are going to plot is contained in a Cricket Graph document, "Example Data."	3
Your overall goal is to create a new graph that matches the example graph shown in the instructions.	4
Your first exercise is to plot the variable *Observed* as a function of the variable *Serial Position.*	5
After you have created a new graph, you will modify it so that it more closely matches the example given in your instructions.	6

model to attend to the correct screen object. Typically, the link between a task goal and the correct object is established through the screen object's label. A label is text that is directly associated with a screen object. Examples include a menu label, a button label, and an icon label.

In summary, Kitajima (1996) demonstrated that the action-planning processes could generate correct action sequences in the absence of device goals, although there are strong constraints on the form of successful task goals. Links between the task goal and the screen object to be acted on at this step must exist, and these links are established by a screen object's label. The words used to describe a task must correspond to the label on the screen object to be acted on. This constraint forces LICAI to predict that label following will be the only exploration strategy available to new users of an application.

3.2. Experimental Task and Instructions

LICAI simulates an experienced Macintosh user's performing the graphing task described in Section 1.2. Our analysis focused on the first task, creating the default graph. Figure 3 shows the instruction texts used in our simulation. These instructions are a simplified version of the instructions used by Franzke (1994), who incorporated more information about the first task. For example, Franzke explicitly stated that the variable *Serial Position* was to be plotted on the x-axis and the variable *Observed* on the y-axis.

In the simulation, each sentence listed in Figure 3 was processed by a single problem-model construction cycle, described in Section 3.4. The numbers in the second column indicate the problem-model construction cycle number. At the end of the integration phase of each cycle, the original propositional representation of each sentence and all propositions generated by the comprehension schemata were transferred to episodic memory, described in Section 3.4. LICAI assumes that the user reads all instructions and then attempts to perform the task described in the text and that the display is the retrieval cue for the task goal that controls action planning for the current step of the task. Successive displays generated by the application serve as retrieval cues for the sequence of task goals.

3.3. Comprehension Schemata

LICAI assumes that goal-formation processes are analogous to Kintsch's (1988) model for solving word problems. This model assumes that reading is a strategic process (van Dijk & Kintsch, 1983) and that strategies generate inferences required to guide problem solving. The knowledge used by the strategies is represented as comprehension schemata. The goal-formation processes in LICAI take a semantic representation of the task instructions as input and elaborate this representation with inferences generated by specialized schemata to construct task goals. The task goals control the action-planning processes that generate exploratory behavior.

LICAI incorporates three kinds of schemata. *Global instruction-reading schemata* represent the top-level strategy used by a reader to process text that describes a given task. All verbs with the implicit subject YOU are mapped into a text-base proposition in the form, DO [YOU, verb, object]. *Task-domain schemata* elaborate DO propositions and generate a more complete description of a task. For example, Franzke (1994) used telegraphic instructions for editing tasks. Editing task-domain schemata elaborate these initial telegraphic descriptions. *Task-goal formation schemata* transform DO propositions into propositions that represent task goals.

The task-domain and task-goal formation schemata are discussed in detail next, including how they elaborate propositional representations of the sentences shown in Figure 3. This text has been transformed into a propositional representation—the text base—using the methods described by Bovair and Kieras (1985). In addition, the global instruction-reading schema has been applied to transform all propositions of form VERB[object] into form DO [YOU, verb, object].

Task-Domain Schemata

Task-domain schemata describe users' specialized knowledge of a task domain that enables them to elaborate the description of the task contained in the text. These schemata represent users' task-formulation knowledge that is independent of a particular application interface. If users have no such knowledge, only the description of the task contained in the text will be transformed into task goals. We have assumed that users have some knowledge of graph and editing tasks. The editing-task knowledge is a generalization of users' experience with word processors.

Example From Data Graph Task Domain

The elaboration processes performed by the task-domain schemata is illustrated by tracing the elaboration of Sentence 5: "Your first exercise is to plot the variable *Observed* as a function of the variable *Serial Position*." The original input sentence is propositionalized as follows:

```
P51   EXERCISE
P52   FIRST [P51]
P53   OBSERVED
P54   ISA [P53, VARIABLE]
P55   SERIAL-POSITION
P56   ISA [P55, VARIABLE]
P57   DO [YOU, PLOT, P58]
P58   AS-A-FUNCTION-OF [P53, P55]
P59   ISA [P57, P51]
```

The above text base is elaborated with the following three task-domain schemata from the data graph task domain:

Plot Schema:

IF ([AS-A-FUNCTION-OF [ARG-1, ARG-2]) →
  [ROLE [ARG-1, DEPENDENT-VARIABLE],
  [ROLE [ARG-2, INDEPENDENT-VARIABLE]

Put Dependent-Variable Schema:

IF ([ROLE [ARG, DEPENDENT-VARIABLE]) →
  ON [ARG, Y-AXIS]

Put Independent-Variable Schema:

IF (ROLE [ARG, INDEPENDENT-VARIABLE]) →
  ON [ARG, X-AXIS]

The meaning of "as a function of" in the original text is elaborated by the plot schema and the two put schemata. Execution of these three task-domain schemata during the construction phase for Sentence 5 adds the following propositions to the network:

```
P60   ROLE [OBSERVED, DEPENDENT-VARIABLE]
P61   ROLE [SERIAL-POSITION, INDEPENDENT-VARIABLE]
```

P62 ON [OBSERVED, Y-AXIS]
P63 ON [SERIAL-POSITION, X-AXIS]

At this stage, Proposition P57 is elaborated by the following:

P57-0 AND [P57-1, P57-2]
P57-1 DO [YOU, PLOT, P62]
P57-2 DO [YOU, PLOT, P63]

Example From Editing Task Domain

The instructions for Franzke's (1994) second task, editing the default graph, were terse descriptions of each edit (e.g., "Change the legend text to Geneva, 9, bold"). This cryptic instruction must be elaborated before the edit task can be understood and executed. For example, "Geneva" must be identified as the new font of the edited legend text. We assume experienced users of word-processing systems have specialized task-domain schemata termed *text-attributes schemata* that transform such terse editing commands into comprehensible instructions. The text base for the original editing instruction follows:

P90 DO [YOU, PERFORM, P91]
P91 AND [P92, P94, P96]
P92 DO [YOU, CHANGE-TO, LEGEND, P93]
P93 PROPERTY [TEXT, $, GENEVA]
P94 DO [YOU, CHANGE-TO, LEGEND, P95]
P95 PROPERTY [TEXT, $, 9]
P96 DO [YOU, CHANGE-TO, LEGEND, P97]
P97 PROPERTY [TEXT, $, BOLD]

The original text does not identify explicitly which text attributes have the values *Geneva, 9,* and *bold.* These inferences are generated by the following text-attributes schema:

Text-Attributes Schema:
IF (PROPERTY [TEXT, $, *ARG*] & ISA [*ARG*, FONT-NAME]) →
   PROPERTY [TEXT, FONT, *ARG*]
IF (PROPERTY [TEXT, $, *ARG*] & ISA [*ARG*, NUMBER]) →
   PROPERTY [TEXT, SIZE, *ARG*]
IF (PROPERTY [TEXT, $, *ARG*] & ISA [*ARG*, STYLE-NAME]) →
   PROPERTY [TEXT, STYLE, *ARG*]

As the results of application of the schema, the original text base is elaborated by the following propositions:

P93-1   PROPERTY [TEXT, FONT, GENEVA]
P95-1   PROPERTY [TEXT, SIZE, 9]
P97-1   PROPERTY [TEXT, STYLE, BOLD]

Task-Goal Formation Schemata

Two task-goal formation schemata generate task goals used by the action-planning processes—*Task* and *Do-It*. These schemata elaborate propositions of the form DO [YOU, *VERB, OBJECT*] in the text base into propositions that represent task goals. The task schema elaborates propositions in the task domain, and the Do-It schema elaborates propositions to perform a specific action on a screen object.

Task Schema

Standard parsing strategies and the global instruction-reading schema generated the following text base for Sentence 1 ("In this experiment, you are going to learn a new Macintosh application, Cricket Graph, by exploration"). The original text base is:

PIO   IN-EXPERIMENT [P17]
PII   YOU
P12   NEW [P13]
P13   MACINTOSH [P14]
P14   APPLICATION
P15   CRICKET-GRAPH
P16   REF [P12, CRICKET-GRAPH]
P17   DO [YOU, LEARN, CRICKET-GRAPH]
P18   BY-EXPLORATION [P17]

By applying task schema to P17, the following propositions are added to the current network:

TASK-10   PERFORM [S10, S11]
S10   TASK-ACTION [LEARN]
S11   TASK-OBJECT [CRICKET-GRAPH]

The propositions TASK-10, S10, and S1 jointly define a task goal.

The task schema requires the *VERB* argument in the proposition DO [YOU, *VERB, OBJECT*] to be a description of TASK-ACTION, like PLOT, CHANGE, and CREATE. The task schema has the following form:

IF (DO [YOU, *VERB, OBJECT*] & ISA [*VERB*, TASK-ACTION]) →
   PERFORM [TASK-ACTION, TASK-OBJECT, TASK-SPECIFICATION],
      TASK-ACTION [*VERB*],
      TASK-OBJECT [*OBJECT*]
      TASK-SPECIFICATION [*list of specifications*]

The arguments in TASK-SPECIFICATION refer to propositions that modify *OBJECT*. The propositions in the consequence part, PERFORM [.], TASK-ACTION[.], TASK-OBJECT[.], and TASK-SPECIFICATION[.], jointly define a task goal for the action-planning processes.

Do-It Schema

In Franzke's (1994, 1995) experiment, participants were given instructions like, "Please click on *General Instructions* to start the experiment." Following such instructions involves significant inferences. For example, "click on" must be mapped onto the action, "Single-click with the mouse button after moving the mouse cursor to the required screen object." *General Instructions* must be mapped onto the screen object with the label *General Instructions*.

When *VERB* in propositions of the form DO [YOU, *VERB, OBJECT*] is a DEVICE-ACTION like click, drag, or release, the do-it schema elaborates such propositions into propositions representing task goals. The do-it schema has the following form, and the arguments in DEVICE-SPECIFICATION refer to propositions that modify the *OBJECT*.

Do-It Schema I:

IF (DO [YOU, *VERB, OBJECT*] & ISA [*VERB*, DEVICE-ACTION]) →
   PERFORM [DEVICE-ACTION, DEVICE-OBJECT, DEVICE-SPECIFICATION]
      DEVICE-ACTION [*VERB*]
      DEVICE-OBJECT [*OBJECT*]
      DEVICE-SPECIFICATION [*list of specifications*]

The text "Please click on *General Instructions* to start the experiment" is propositionalized as follows:

P20 EXPERIMENT
P21 $
P22 HAS-LABEL [P21, General Instructions]
P23 DO [YOU, CLICK, P21]
P24 BY [P25, P23]
P25 DO [YOU, START, EXPERIMENT]

In P23, CLICK is a kind of DEVICE-ACTION; thus, the do-it schema generates the following propositions representing a task goal:

        PERFORM [S21, S22, S23],
  S21 DEVICE-ACTION [CLICK],
  S22 DEVICE-OBJECT [$]; undefined
  S23 DEVICE-SPECIFICATION [DEVICE-LABEL [General Instructions]]

There are cases in which "click on" may be replaced with a representation of general actions (e.g., "act on") that do not directly indicate device actions, whereas a screen object is indicated. Another version of the do-it schema exists for these cases. In the example instruction, "Example Data" in Sentence 3 can trigger the rule:

Do-It Schema II:

IF (HAS-LABEL [*O BJEC T, LABEL*]) →
    PERFORM [$, DEVICE-OBJECT, DEVICE-SPECIFICATION]
    DEVICE-OBJECT [*O BJEC T*]
    DEVICE-SPECIFICATION [DEVICE-LABEL [*LABEL*], *list of specifications*]

3.4. Reading Instructions

The simulation of reading the instruction text (provided in Figure 3), generating task goals (and then storing them), and carrying over a few nodes entails processing the text sentence by sentence in a series of problem-model construction cycles. For each sentence, the comprehension schemata defined in the previous section are applied to generate an elaborated text base of the original sentence. The program developed by Mross and Roberts (1992) simulates the C–I cycles for the elaborated text base of each sentence. The simulation of reading the first sentence in Figure 3 is described in the following three subsections; the result of reading the whole instructions is presented in the fourth subsection.

Construction and Integration of the Network

As described in Section 3.3, the text base representing Sentence 1 consists of 9 propositions, P10 to P18. and 3 additional propositions generated by the task schema. Figure 4 illustrates these results. The oval nodes represent the original text, and the output of the task schema is depicted as rectangles. These propositions are connected by their shared arguments. For example, P14: APPLICATION and P13: MACINTOSH[P14] (shorthand for MACINTOSH[APPLICATION]) are linked by the shared argument, APPLICATION. The

Figure 4. A network representation for the first sentence, "In this experiment, you are going to learn a new Macintosh application, Cricket Graph, by exploration."

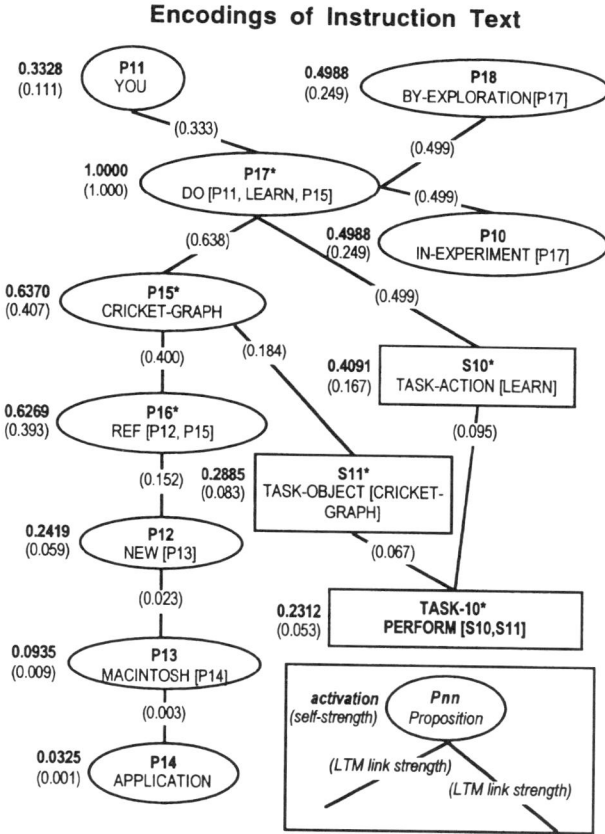

Encodings of Instruction Text

simulation program reads the list of propositions as input and then constructs the linked network, represented as a matrix, **C**. The initial values for the link strengths are set to 1.0.

The network is then integrated by the following procedure. $\mathbf{A}^{(0)}$ is a row vector that represents the activation values and that has the same size as matrix **C**. Initially, the activation values for the source nodes, P10 to P18, are set to 1, and the activation values for the elaborations are set to 0, $\mathbf{A}^{(0)} = (1, 1, 1, 1, 1, 1, 1, 1, 1, 0, 0, 0)$. The activation vector for the first iteration of the integration cycle $\mathbf{A}^{(1)}$ is computed by first postmultiplying $\mathbf{A}^{(0)}$ by **C**—that is, $\mathbf{A}^{(0)} \times \mathbf{C}$—and then normalizing the resulting vector. The normalization process sets negative activation values to 0 and then scales the positive values so that the activation value of the most highly activated node equals 1.0. The iteration process continues until the network converges—namely, until the average change in the activation vectors for two successive iterations becomes less than an arbitrarily determined threshold value (e.g., 0.001). The activation vector for the last iteration represents the outcome of the comprehension process. The boldface numbers next to the nodes in Figure 4 are activation values for the network that reached convergence after 11 iterations.

Forming Episodic Memory

When the network is integrated, the results are transferred to episodic memory. The process, based on Kintsch and Welsch's (1991) model, is carried out automatically by Mross and Roberts's (1992) program. The episodic memory is represented as a collection of propositions—the unique propositions generated during the problem-model construction cycles. In our example, when the first cycle is completed, the episodic memory is empty; thus, copies of nodes in the working memory shown in Figure 4 comprise episodic memory. Each node in episodic memory has a "self strength," reflecting the activation value of the corresponding node in the integrated network. Likewise, each link in episodic memory has a "link strength," reflecting the activation values of the corresponding nodes in the integrated network. Figure 4 shows the values of self strengths and link strengths in parentheses as calculated by Mross and Roberts's program.

Carrying Over Propositions to Maintain Coherence

The purpose of the sequence of problem-model construction cycles is to generate a representation of the text, not just individual sentences. Linking the representations of individual sentences establishes textual coherence. In addition, the model takes advantage of its understanding of the instruction that has been achieved so far. The process of establishing textual coherence and accumulating knowledge about the task is modeled by carrying over propositions from one cycle to another. The following carry-over process was used by Kintsch (1988).

After storing the result of comprehension of Sentence n and before processing Sentence $n + 1$, the model carries over a few nodes in working memory to maintain coherence and accumulate knowledge about the task. Textual coherence is maintained by carrying over the three most highly activated nodes to the next processing cycle. Accumulation of knowledge about the task is accomplished by carrying over a set of propositions that originated from a task-goal formation schema. For example, after processing Sentence 1, the program carries over PI7, PI5, and PI6 and TASK-10, SI0, and SII to the cycle for Sentence 2.

Figure 5. **Task goals generated from the example instructions by the problem-model construction cycles.**

Sentence	Proposition Label	Task Goal (PERFORM [ACTION, OBJECT, SPECIFICATION])
1	TASK-10	PERFORM [LEARN, CRICKET-GRAPH]
2	TASK-20	PERFORM [PERFORM, TASK]
3	TASK-30	PERFORM [PLOT, DATA]
3	DO-IT-31	PERFORM [$, $, SPEC [LABEL [EXAMPLE-DATA], DOCUMENT]]
4, 6	TASK-40	PERFORM [CREATE, GRAPH]
5	TASK-60	PERFORM [PLOT, $, AS-A-FUNCTION-OF [OBSERVED, SERIAL-POSITION]]
5	TASK-61	PERFORM [PLOT, OBSERVED, Y-AXIS]
5	TASK-62	PERFORM [PLOT, SERIAL-POSITION, X-AXIS]
6	TASK-70	PERFORM [MODIFY, GRAPH]

Results of Problem-Model Construction Cycles

The six sentences shown in Figure 3 were processed by a sequence of problem-model construction cycles, and Figure 5 shows task-goal propositions generated by these cycles.

3.5. Exploration

After processing the instructions, LICAI enters the goal-selection and action-planning processes that generate exploratory behavior. These processes involve a memory-retrieval cycle followed by one or more attention–action-selection cycles. The memory-retrieval cycle uses display as a retrieval cue to retrieve a task goal. The retrieved task goal is passed to Kitajima and Polson's (1995) action-planning model, operating without device goals, to generate one or more actions. The resulting display is used as a new cue, and the process continues. LICAI simulated the first three steps of Franzke's (1994) first task, and Figure 6 shows the initial sequence of displays generated during exploration.

Goal-Selection Processes

Each display contains a collection of screen objects, and each screen object is represented as a few propositions (see Kitajima & Polson, 1995, for details). In the simulation of the memory-retrieval cycle, propositions representing screen objects that link task goals in episodic memory were incorporated into the network with all of the task goals. The memory-retrieval cycle was simulated using Mross and Roberts's (1992) program.

Figure 7 is a schematic representation of the network for the memory-retrieval cycles. The display cues are represented as a set of propositions that convey perceptual information, such as the label of a screen object

Figure 6. Description of displays used for the simulation of goal-selection processes. Note that part of each display was used for retrieving a task goal from the episodic memory. Versions of displays in Franzke's (1994) experiment are shown for reference.

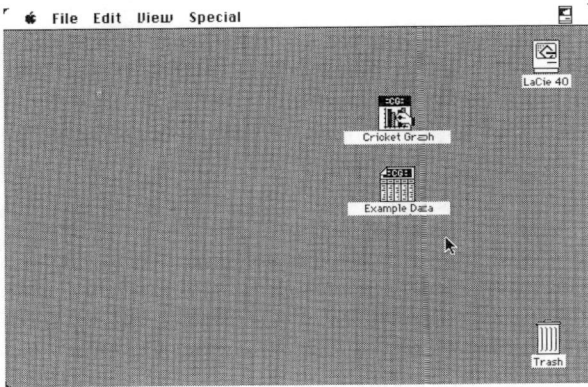

Display I = beginning of task. Simulation of goal-selection process included:

- Two desktop icons, *Cricket Graph* and *Example Data.*

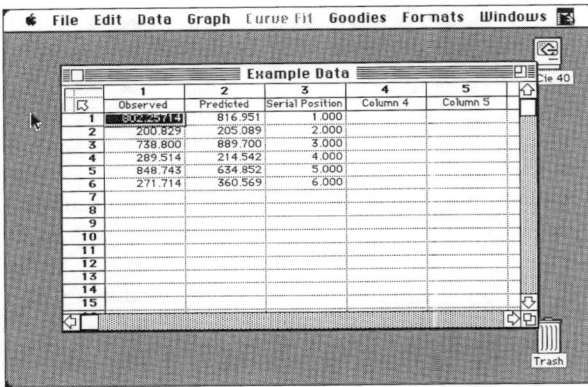

Display II = after launch of Cricket Graph I by opening *Example Data* document icon. Simulation of goal-selection process included:

- Two items from the menu bar, *Data* and *Graph.*
- Labels from spreadsheet, *Serial Position* and *Observed.*

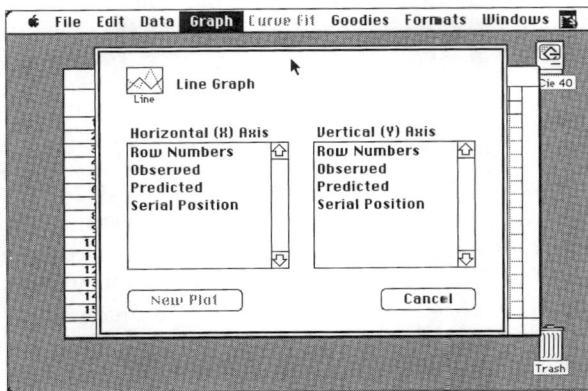

Display III = variable-selection dialog box for assigning variables to x-axis and y-axis obtained by selecting *Line* from *Graph* menu. Simulation of goal-selection process included:

- Two pairs of *Serial Position* and *Observed* in the variables-selection dialog box.

Figure 7. Retrieval of task goals, **PERFORM [ACTION, OBJECT, SPEC]**, cued by external screen representations.

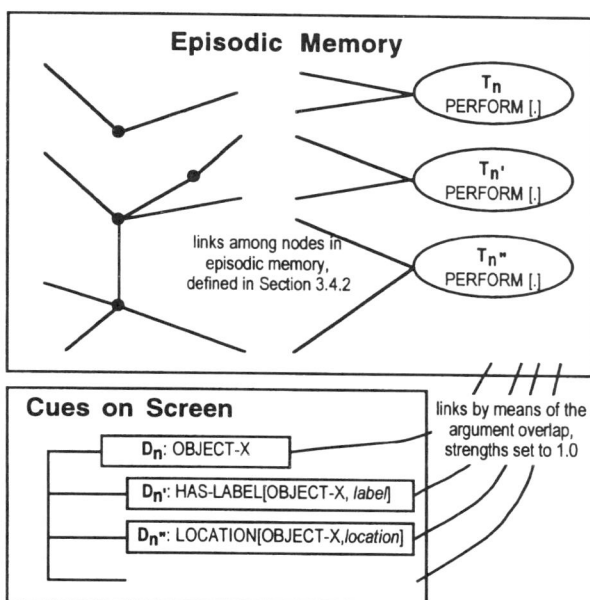

and its state (e.g., highlighted). For example, Display I (shown in Figure 6) is represented as seven propositions:

```
D-10    OBJECT-I0
D-I0I   HAS-LABEL [OBJECT-I0, CRICKET-GRAPH]
D-I02   ISA [OBJECT-I0, APPLICATION]
D-20    OBJECT-20
D-20I   HAS-LABEL [OBJECT-20, EXAMPLE-DATA]
D-202   ISA [OBJECT-20, DOCUMENT]
D-203   CONTAIN [OBJECT-20, DATA]
```

Links are formed between display propositions and propositions in episodic memory that share arguments. These links, defined by overlapping arguments, are given a strength of 1.0. In the network-integration process, the node representing the display object itself, D_n, serves as a permanent activation source whose activation value is always reset to 1.0 before an iteration of the spreading activation process. Other nodes in the display representation are initially set to 1.0, but their activation values can change on each iteration. The nodes in episodic memory are set to their self strengths as their initial activation values. At the end of the integration phase of the memory-retrieval cycle, the most highly activated task goal is passed to the action-planning processes.

Figure 8. Activation values of potential task goals after a memory-retrieval cycle (unit 0.0001).

Proposition Label and Task Goal	Display I	Display II	Display III
TASK-10: PERFORM [LEARN, CRICKET-GRAPH]	19	0	0
TASK-20: PERFORM [PERFORM, TASK]	0	0	1
TASK-30: PERFORM [PLOT, DATA]	8	3	0
DO-IT-31: PERFORM [$ $ SPEC [LABEL [EXAMPLE-DATA], DOCUMENT]]	51	0	0
TASK-40: PERFORM [CREATE, GRAPH]	3	39	14
TASK-60: PERFORM [PLOT, $, AS-A-FUNCTION-OF [OBSERVED, SERIAL-POSITION]]	0	1	4
TASK-61: PERFORM [PLOT, OBSERVED, Y-AXIS]	0	4	21
TASK-62: PERFORM [PLOT, SERIAL-POSITION, X-AXIS]	0	4	21
TASK-70: PERFORM [MODIFY, GRAPH]	0	19	7

Figure 8 shows the results of the goal-selection processes by the activation values for each task goal proposition on each of the three display conditions. The actual activation values of each task goal proposition can be calculated by multiplying each entry by 0.0001.

Action-Planning Processes

Simulation of the action-planning processes was performed by examining the selected task goal in the light of the result by Kitajima (1996) to predict the anticipated result of the action-planning processes.

The correct action for the initial step of the first task is to move the mouse cursor to the document icon labeled *Example Data* and double-click on it to open the document. Display I (shown in Figure 6) was the retrieval cue. As shown in Figure 8, the most highly activated task goal was PERFORM [$, $, SPEC [LABEL [EXAMPLE-DATA], DOCUMENT]], DO-IT-31. This proposition represents the task goal of performing a yet to be specified action on an unspecified screen object—a document labeled *Example Data*. The argument EXAMPLE-DATA overlaps with the label on the correct document icon in Display I. In addition, the argument DOCUMENT reinforces the link between the task goal and the correct screen object. This link is through the proposition D-202: ISA [OBJECT-20, DOCUMENT]. The argument DOCUMENT had significant effect on the resulting activation pattern. Unless DOCUMENT were specified in DO-IT-31, a competing task goal, TASK-10, would have been equally activated as DO-IT-31. These links enable the attention cycle to select the correct screen object.

The correct action in the second step is to pull down the *Graph* menu. LICAI retrieved TASK-40, PERFORM [CREATE, GRAPH] using Display II as a retrieval cue. The link between the screen-object label representing the *Graph* menu and the GRAPH in TASK-40 caused this task goal to become the most highly activated. This same link causes the action-planning processes to attend to the *Graph* menu item, move the cursor to it, and pull it down.

The links from Display III to TASK-61, PERFORM [PLOT, OBSERVED, Y-AXIS] and TASK-62, PERFORM [PLOT, SERIAL-POSITION, X-AXIS] caused both task goals to become equally highly activated. There are multiple links from screen objects that make up the variable-selection dialog boxes to these propositions. However, there is no mechanism in LICAI to resolve a tie. Kitajima (1996) showed that each goal would cause the action-planning processes to select the correct variable name in each dialog box. Thus, in this case, each task goal could generate correct actions regardless of the order in which the highest activated task goals were passed to the action-planning processes.

Summary

These simulation results clearly demonstrate LICAI's dependence on various manifestations of the label-following strategy in the exploration processes. Links between task goals and object labels on the display play identical key roles in mediating correct performance. Labels for the correct screen objects form links that control successful retrieval of the correct task goal and that guide successful action planning. In addition, Kitajima and Polson (1997) showed that these same links are involved in successful retrieval of action sequences.

4. EVALUATION OF LICAI

We evaluated LICAI using Franzke's (1994, 1995) results. The model only makes coarse predictions about the behavior of Franzke's participants. The instructions and comprehension schemata assumed by LICAI might enable the model to generate correct task goals, but the model does not describe the search behavior that occurs if the task-goal construction process fails (Rieman et al., 1996). In fact, Rieman et al. (1996) found that people will do some exploration of the interface even when they can construct the correct task goal and execute the correct action within 15 to 30 sec.

4.1. Task Descriptions, Object Labels, and Number of Screen Objects

Franzke (1995) presented an analysis in which she collapsed her data across interfaces and tasks. She focused on the relations among task descriptions for each step, the correct object labels, and the number of screen objects that were possible targets. She analyzed the time it took participants to complete a step. If participants failed to discover the correct action within 2 min, she gave a hint.

Franzke (1995) partitioned the possible relations between a task description and the label of the correct screen object into four categories: (a) a perfect match between the label and the task descriptions, (b) the labels were synonyms of the task description, (c) an inference was required to link the two, and (d) no label on the object. All direct manipulation actions (like double-click to gain access to an editing dialog box) were in the fourth category. Franzke also used a coarser classification (*good, poor*) in which the first two categories are the *good* labels. We did not simulate all of the tasks and interfaces used in Franzke's experiments because our goal was to demonstrate that LICAI can provide a qualitative account of Franzke's major results. Recall that LICAI will fail to discover the correct action if there is not a link from the task description to the correct screen object.

Label Following

Franzke (1995, Figure 5) found strong support for the label-following strategy (Polson & Lewis, 1990; Polson et al., 1992). The time required to perform the correct action was well under 30 sec for the match (a) and synonym (b) categories, and the time required for the other two categories (c and d) was significantly longer. In fact, a large fraction of the no-label (d) relation steps required hints. During exploration, participants used overlap between task descriptions contained in the instructions and labels on menus, buttons, and other interface objects.

The degree of success on first attempt at a task was also dependent on the number of screen objects. Franzke (1995, Figure 6) showed that the more objects displayed, the longer participants took to complete the step. Franzke (1995, Figure 7) also analyzed action times for discovering correct actions and found an interaction between number of objects (2 to 10) on the screen and quality of the label match (*good* or *poor*). There was no effect of number of objects for *good* labels and a large effect for *poor* labels. A unique *good* label caused a person to attend to the correct screen object regardless of the number of screen objects.

These results are consistent with LICAI. If an instruction contained a description of either an action or an object that matched a screen-object label, those labels are preserved when the propositional representation of the instruction is mapped into a task goal. The links between the task goal and the correct screen object can mediate performance of the correct action if the matching label is unique, even when there is a large number of screen objects. The correct screen object will be included in the set of candidate objects in the attention cycle because the unique links established by the matching label will cause selective activation of the correct node.

We did not simulate the success of synonyms in mediating successful label following. However, the result are consistent with the C–I architecture. Common synonyms would be retrieved during the elaboration process and added to the network. These synonyms would then enable the model to construct links between task goals and plan elements required in the action-planning processes.

Direct Manipulation

Franzke (1995, Figure 8) found that participants had difficulty discovering direct-manipulation actions when there were no label links between the correct screen object and the task description. Examples included double-clicking on a title or an axis label to edit it, clicking on an object in the tool bar, and dragging and dropping items. In a majority of first encounters with an example of such interactions, Franzke's (1994) participants had to be provided hints after 2 min of futile exploration. Franzke (1995) argued that participants did not have the knowledge necessary to infer that an object could be edited by double-clicking on it. Thus, participants could not perform such tasks even if they had generated the correct task goals from the instructions. LICAI predicts such failures because links between the task goal inferred from the instructions and the correct screen object are missing.

4.2. Franzke's (1994) Instruction Texts

Franzke (1994) used two types of instructions in her experiments. The initial instructions enabled participants to generate the default graph and were more detailed than the instructional text used in our simulation. Participants could easily generate multiple task goals from such texts. The texts that described the edits to be performed on the default graph, on the other hand, were short and telegraphic. People's comprehension problem here was to make the inferences necessary to generate an understandable task description and construct task goals that overlapped with screen-object labels.

Task-Domain Schemata

The critical step in performing the first task, creating the default graph, is interacting with the dialog box, which contains two scrolling lists for the x- and y-axis variable labels. LICAI transformed the original problem statement, "Plot *Observed* as a function of *Serial Position*," into "Plot *Observed* on the y-axis" and "Plot *Serial Position* on the x-axis" by using task-domain schemata (described in Section 3.3).

When Terwilliger and Polson (1997) measured the time that experienced Macintosh users, who had never used a graphing application, took to interact with two forms of the variable-selection dialog box, they found clear evidence that people also make such transformations. Terwilliger and Polson constructed two versions of task instructions for Franzke's first subtask. The "XY" instructions read, "Create a graph with *Serial Position* on the x-axis and *Observed* on the y-axis"; the "FN" instructions read, "Create a graph of *Observed* as a function of *Serial Position*." Terwilliger and Polson also created two versions of the dialog box. In the XY version, the left selection list was labeled *X Axis,* and the right selection list was labeled *Y Axis;* in the FN version, the left list was labeled *Plot,* and the right list was labeled *As a function of.* Terwilliger and Polson found that tasks took less time with the XY dialog box for XY and FN instructions. In addition, the think-aloud protocols recorded while participants performed the task revealed that they verbalized the transformation of FN to XY. These results support the concept of task-domain schemata as well as the assumption that the label on the interface must match user descriptions of a task.

To perform the second task, "Change the legend text to *Geneva, 9, bold,*" participants had to first double-click on the legend to open a dialog box. Most participants had trouble discovering this initial action and had to be given hints. After the dialog box was open, however, participants had no trouble completing the task. The dialog box contained three scrolling lists labeled *Font, Size,* and *Style.* LICAI also generated the three task goals necessary to interact with the three scrolling lists (see Section 3.3) and performed each task specified by the original text. The text-attributes schema had generated the task goals that link to a scrolling list title and to the relevant item in the scrolling list in the dialog box.

Long Instructions and Multiple Task Goals

Franzke's (1994) instructions for the first task, creating the default graph, were more detailed than the instructions in our simulation and included general information on the experiment as well as details about which variables to place on the x- and y-axes. Even so, participants in the Cricket Graph I and III conditions still had some difficulty with these initial steps and had to reread the instructions and obtain critical information like variable names. Their difficulty is somewhat surprising, because the instructions used terms that provided a perfect match for the label-following strategy. As LICAI predicts, however, these instructions would generate multiple task goals, and retrieval of these goals is a brittle process, heavily dependent on the specific display representation.

4.3. Summary

We have shown that LICAI accounts for the major results of Franzke's (1994, 1995) extensive study of learning by exploration. A good match between one or more elements of a user's task and a label on a menu, button, or other screen object can guide successful exploration. LICAI's goal-selection and exploration processes are completely dependent on such overlaps. Thus, our model provides a good account of Franzke's results.

The goal-formation processes are more complex. Various comprehension schemata can transform the original problem statement in complex ways. Thus, the problem statement, "Plot *Observed* as a function of *Serial Position*," is transformed into "Put *Observed* on the y-axis" and "Put *Serial Position* on the x-axis." Kitajima (1996) showed that this transformation is necessary for LICAI to be able to perform the subtask of generating the default graph by exploration. Terwilliger and Polson (1997) found direct evidence that users will transform a problem statement of the form "Plot ... as a function of ..." in the manner consistent with LICAI's task-domain schemata.

However, LICAI will map the problem statement in the text almost directly into a task-goal representation in the absence of any domain-specific comprehension schemata that could transform or elaborate the original text, generating multiple task goals. Selecting a correct task goal could become a difficult task.

5. MODELS OF EXPLORATION

This section compares LICAI—a comprehension-based model of exploration—with another class of models of exploration, IDXL (Iteratively Deepening eXploratory Learner; Rieman et al., 1996), a search-based model of exploration build on the Soar architecture. The key difference is that comprehension is a highly automated collection of cognitive processes that uses large amount of knowledge, whereas search-based models use mechanisms that enable them to deliberately search the interface and memory for relevant cues and linking knowledge. Section 5.1 characterizes the comprehension-based models. Section 5.2 describes three differences between search-based models and LICAI.

5.1. Comprehension-Based Models

Kintsch's (1988) and Kintsch and Greeno's (1985) models of arithmetic word problems are instances of a general class of models that have been proposed repeatedly in the literature on problem solving and skill acquisition (Greeno & Simon, 1988). These models have a comprehension-based problem-representation building component and a problem solver that ultimately generates the solution. An early example of this class was Simon and Hayes' (1976) UNDERSTAND model, which processed instructions for tasks like the tower of Hanoi and generated a representation that was input to the general problem solver (Ernst & Newell, 1969).

LICAI and Kintsch's models of arithmetic word problems are extreme versions of this class of models in that they have no problem-solving mechanisms. Many of the cognitive operations that other investigators would characterize as problem-solving activities are instead done in the comprehension component. This difference is justified by Kintsch (1988), who argued that children's difficulty with arithmetic word problems is not in performing the basic arithmetic operations but in comprehending the problem description, and by Hudson (1983), who showed that linguistic factors can have a powerful effect on problem-solving success.

Because Kitajima and Polson's (1995) action-planning model requires explicit task goals to generate correct actions, the form of useful goals is strongly constrained by the surface features of the interface—like labels in dialog boxes and on menus. Assuming that skilled users have specialized comprehension knowledge directly analogous to Kintsch and Greeno's (1985) arithmetic schemata, the schemata should enable them to construct the required specific goal representation from text descriptions of their task. On a more general level, the flexibility and power of an expert user's skills in using a specific application are in the problem-formation component and not in the action-planning component.

5.2. Search-Based Models and LICAI

The C–I theory was developed to account for reading, a highly skilled behavior. It was then extended to account for action planning in skilled computer users (Kitajima & Polson, 1995; Mannes & Kintsch, 1991). In contrast to the ACT–R (Anderson, 1993) or Soar (Newell, 1990) architectures, C–I theory has a primitive control structure, but it does have components that enable it to exhibit search behavior, including goals, a mechanism to choose between alternative actions, and the ability to react to results of a selected action.

LICAI starts in comprehension mode by reading the instructions and storing alternative task goals in memory. It then switches to action planning and attempts to execute the correct actions by using cues generated by successive displays to retrieve task goals from memory. LICAI, however, cannot interleave comprehension and action planning. The underlying C–I architecture does not support the structures necessary to pass control between action-planning and comprehension modes based on the current state of comprehension or action-planning processes.

Alternative theories of exploration do have such control structures, however. IDXL, a Soar model developed by Rieman et al. (1996), simulates learning by exploration of the Cricket Graph task and integrates much recent research on learning by exploration (Howes, 1994; Howes & Young, 1996; Rieman, 1994; Rieman et al., 1994).

There are three important differences between LICAI and IDXL. One is grain size. IDXL accounts for the actual search behavior by modeling the user's scanning of the display and examination of pull-down menus. Rieman (1994) conducted a detailed analysis of the actual exploratory behavior of those Franzke and Rieman (1993) participants who learned Cricket Graph I and found that participants will explore the display even when a menu with a label matching the current goal is present. Participants exhibit a form of "iteratively deepening attention" (Rieman et al., 1996). The following menu-scanning behavior is an illustration. On first encounter, a menu is pulled down and quickly scanned, and then the mouse cursor is moved to another screen object. During a later encounter, participants study menu items by highlighting and pausing on each one in succession.

IDXL has a primitive model of attention that focuses on one screen object at a time. The action-planning operations include scanning, pointing, dropping pull-down menus, releasing on a menu item, and the like. The comprehension processes operate on the current attended-to screen object and compare its label to a current task goal. The output of the comprehension operators includes forming an exact match, recognizing a synonym, recalling, constructing an analogy, asking for instruction, and envisioning consequences.

LICAI's action-planning processes model the same behavior but at a more abstract level. The model retrieves a single task goal, considers all screen objects concurrently, and generates an action sequence associated with the task goal. IDXL can account for situations in which there is not a perfect match between the current task goal and the correct screen-object label; the user discovers the correct action after a significant amount of search. LICAI simply fails to generate the correct action sequence.

A second difference is that IDXL interleaves comprehension and action planning. Both are forms of progressively deepening search mechanisms, and both operators are ordered by cost. Scanning actions (e.g., moving the mouse cursor; pulling down a menu) have low cost, but releasing on an incorrect menu item can be costly. Comprehension operators are applied to screen objects put in the focus of attention by scanning operators. Initially, low-cost comprehension operators are applied like a test for an exact match to the current task goal. Failure causes IDXL to continue scanning. When the model's attention returns to the same screen object, results of the previous comprehension operators are recalled and lead to application of more costly comprehension operators. IDXL will act when the representation of the proposed action, generated by successive application of more costly comprehension operators, has generated a good match to the current task goal.

The third difference is that IDXL does not comprehend the original task instructions. IDXL is supplied with a task description in working memory that is a multipart, hierarchical representation in a subject–verb–object format. IDXL assumes a shifting internal focus of attention, although it has not been modeled yet, to define a current task goal (Rieman et al., 1996).

It is clear that developments of LICAI must extend the underlying architecture with control mechanisms that enable the interleaving of comprehension and search. A principled account of the search behavior described by Rieman (1994) and simulated by IDXL must be developed. However, the resulting model would be different from IDXL. IDXL's comprehension operators elaborate the user's representation of objects on the interface with a fixed representation of a current task goal. These operators are similar to the elaboration phase in Kitajima and Polson's (1995) action-planning model. An extension of LICAI, in contrast, would be to search for alternative interpretations of the instructions that enable the action-planning mechanisms to make progress on the task.

Franzke's (1994) instructions for the task simulated by IDXL were similar to but more detailed than the instruction input to LICAI. In both cases, the comprehension processes generated multiple task goals. Our interpretation of much of the search behavior observed by Franzke and Rieman (1993) is that Franzke's participants were searching for possible interpretations of the instructions to find useful task goals. LICAI simulates a participant who processes all instructions. However, Simon and Hayes (1976) observed that participants have a strong tendency to skim instructions and that they attempt to solve the problem without all of the necessary information. They are then forced to return to the instructions, often several times, to acquire the necessary knowledge. Thus, part of the search behavior described by Franzke and Rieman is caused by incomplete and/or multiple interpretations of the instructions.

6. DISCUSSION

We have developed and evaluated LICAI—a model of display-based HCI that has goal-formation and action-planning processes, both based on Kintsch's C–I architecture. The goal-formation processes transform initial task descriptions into goals that drive the action-planning processes and that are specialized comprehension strategies using comprehension schemata to construct the required goal.

6.1. Comprehension Schemata for Instruction Taking

The schemata, triggered by specific features of the textbase, add specialized elaborations and inferences to the network during the construction phase of each cycle. Two schemata are defined in terms of their content. The task-domain schemata elaborate the original statement of the task description. Examples include plot, put dependent variable, put independent variable (see Section 3.3), and text-attributes schema (see Section 3.3). These schemata describe the user's specialized knowledge of the task domain, and they are independent of the application interface. Task-goal formation schemata (task schema and do-it schema) generate goals used by the action-planning processes. LICAI claims that instruction reading is a strategic process that uses these schemata. The plot and put dependent-variable and put independent-variable schemata are supported by Terwilliger and Polson's (1997) results.

6.2. Label Following

Label following is simulated in LICAI by links between arguments in propositions representing goals and propositions representing the labels on screen objects (e.g., menu items and button labels). As Kitajima's (1996) simulation showed, behavior of the action-planning processes is fragile when there is no device goal. New users of an application cannot generate device goals, and LICAI predicts that label following is the only strategy for successful exploration.

LICAI also predicts that successful exploration will occur only under circumstances in which the instruction-comprehension processes generate task goals that satisfy the label-following constraint. However, the labels that define task goals are not necessarily present in the original instruction texts. The original instruction texts can be elaborated by task-domain schemata, such as the text-attributes schema.

6.3. Multiple Task Goals

Another fundamental result dictated by the underlying architecture is that users studying instructions or an example will generate multiple task goals. Because the construction process is bottom-up, there is no control process to force the model to incrementally generate a single task goal to be acted on after the instructions have been processed. Kintsch (1988) showed that selecting the correct problem model for arithmetic and algebra word problems was facilitated by both the situation context described in the problem text and the last question sentence of the problem description. For computer users interacting with a novel application, successive displays generated by the application serve as retrieval cues for task goals constructed during instruction processing. There are no control processes that generate the proper sequence of correct task goals.

6.4. Implications for Learning by Exploration

Our results have important implications for the development of training materials for interfaces that support learning by exploration (Wharton et al., 1994). A user's background knowledge constrains the content of instructional materials, and the interfaces that support learning by exploration and the instructions that intend to facilitate this form of learning are dependent on one another.

The Minimalist Instruction Paradigm

Carroll (1990) summarized an influential research program on the design and evaluation of training materials for application programs. This work led to the development of the "minimalist instruction paradigm," which was stimulated by the then surprising result that carefully designed, detailed training and reference materials for early 1980s word processors were unusable (e.g., Mack, Lewis, & Carroll, 1983). Carroll's (1990) solution to the usability problems demonstrated by Mack et al. (1983) and other researchers was to refine the minimalist instruction paradigm. The minimalist approach focuses on users' tasks rather than on describing a system function by function, minimizes the amount of written materials, and tries to foster learning by exploration rather than provide step-by-step instructions. This approach also supports error recognition and recovery.

Carroll's (1990) emphasis on active users' desires to learn by exploration are supported by Rieman's (1994) results. However, LICAI shows that following written instructions is complex, analogous to solving arithmetic and algebra word problems. Successful instruction following requires application of the appropriate schemata to extract information from the text to guide generation of correct task goals. Thus, for a new user, following even the most carefully prepared instructions is a difficult and error-prone process.

In addition, LICAI predicts that comprehending and following instructions becomes more difficult, especially for individuals with limited background knowledge. Mack et al.'s (1983) participants were new users and therefore did not have the necessary schemata or action-planning knowledge assumed by LICAI. Thus, instructions for new users must either be detailed or incorporate extensive pretraining (like the "Guided Tour"[5] for the Macintosh interface) to explain how to use the mouse, select items from menus, and so forth. All this knowledge is incorporated in the action-planning component of LICAI.

LICAI can be used to develop explicit design guidelines for the content of minimalist instruction materials because the paradigm and the model both focus on the users' tasks. Task goals, representing what the user wants to achieve, drive the action-planning processes. Explicit, brief statements of relevant goals that can be understood by the user are critical ingredients in supporting learning by exploration. Many irrelevant task goals generated during the comprehension of the example instructions were caused

---

5. Guided Tour is a tutorial program that comes with Mac OS 7.0 and that introduces users to the basic Macintosh interface conventions.

by attempts to provide some motivation for the user's task. Carroll explicitly recommends deleting such irrelevant material.

LICAI also clarifies the constraints that must be understood to follow Carroll's (1990) design heuristic of minimizing the amount of written material contained in task instructions. Brief instructions solve the problem of dealing with multiple task goals, but comprehension of terse instructions, like those used by Franzke (1994), requires specialized background knowledge of the task and interface. For example, the task description, "Change the legend text to *Geneva, 9, bold*," cannot be understood by someone who has no experience with a word processor. Effectively minimizing the amount of written material requires careful attention to the action and display knowledge and the schemata assumed in the target population. A minimalist version of an instruction manual for word processors would need to assume that users have well-developed versions of task and do-it schemata. These schemata represent significant background knowledge (i.e., at least 6 months of experience).

Carroll's (1990) minimalist instruction paradigm focuses almost exclusively on training materials. However, there are important constraints that must be satisfied by the interface for learning by exploration to even be possible: The label-following strategy must work. LICAI clarifies the intimate relation between the content of minimalist instructions and the details of user interfaces that support learning by exploration. The key is the label-following strategy. The model predicts that successful minimalist instructions must guide users to form task goals that contain terms that overlap with the labels of the correct screen objects. Such task goals enable the action-planning processes to select the correct action or action sequence without any previous instruction and without detailed, step-by-step guidance. If the interface does not support label following, then the document designer must use step-by-step instructions, although Carroll (1990) and numerous other investigators have shown that people are reluctant, if not unwilling, to read and follow such instructions.

Cognitive Walkthroughs

Cognitive walkthrough evaluates interfaces for ease of learning by exploration. The walkthrough is organized like the structured walkthroughs widely used in the software development community (Yourdon, 1989) and is based on an earlier theory of exploratory learning (Polson et al., 1992). This method evaluates the effectiveness of the label-following strategy and characterizes the background knowledge necessary to infer correct actions (Wharton et al., 1994). For each action required to perform a task, a designer must show that users have a correct goal that guides selection of the next correct action. The label-following strategy is the primary action-selection guide. The method also forces designers to specify the task and interface knowledge required by users to infer correct actions. For example, it is possible to select an object in a scrolling list by single-clicking on it.

The theoretical analysis presented in this article reinforces the importance of the label-following strategy. In addition, it points to some limitations of the cognitive-walkthrough method. A majority of the successful applications of the cognitive walkthrough has been on walk-up-and-use interfaces like automated teller machines or phone-based applications with voice menus. In all of these situations, there are a limited number of isolated tasks, and the interface guides the user through the procedure necessary to accomplish each task. The instruction for each step is brief and contains only the information necessary for the user to select the next action. Thus, the interface controls the processes involved in interleaving goal formation and action planning. More complicated tasks involving document preparation or the generation of data graphs give the user far more degrees of freedom and are more complex as a result.

In many instructional situations, users are given instructions of a paragraph or two similar in form and content to the example used in our simulations. LICAI generates multiple task goals when processing such instructions. Cues provided by the interface must control the retrieval processes to generate the proper sequence of task goals. These retrieval processes are not robust. A minor change in wording in instructions could lead to the retrieval of incorrect task goals.

7. CONCLUSIONS

This article has presented a model of exploration based on a cognitive architecture, the C–I framework, that has no detailed model of deliberate cognition; it cannot engage in serious search behavior. In spite of the model's limitations, it provides insight into the difficulties individuals have in comprehending detailed technical documentation. The model also provides insights on the strengths and limitations of the minimalist instruction paradigm as well as the cognitive-walkthrough interface evaluation procedure.

Another option that we explored and rejected was to extend the C–I architecture with a Soar-like control structure to support sophisticated search. Living within the limitations of the C–I architecture, we were forced to focus on comprehension and the processes of goal formation and showed the background knowledge necessary to generate the highly constrained goals required by the model. LICAI is closely related to successful models of arithmetic word problem solving (Kintsch, 1988), and the major thrust of both models is that following instructions to carry out some procedure is a difficult task and that comprehending instructions requires specific, specialized background knowledge.

NOTES

Background. This article is an extended version of Kitajima and Polson (1996) and includes new simulation results.

Acknowledgments. Richard Young provided a detailed review of a much longer, early version of this article, and he gave us excellent guidance in his role as action editor. We thank him for helping us clarify many issues and shorten the article. John Rieman, Clayton Lewis, and Walter Kintsch also made important contributions. We also thank Suzanne Mannes and two anonymous reviews for their criticisms and suggestions.

Support. We gratefully acknowledge research support under National Aeronautics and Space Administration (NASA) Grant NCC 2–904. The opinions expressed in this article are those of the authors and are not necessarily those of NASA.

Authors' Present Addresses. Muneo Kitajima, National Institute of Bioscience and Human-Technology, 1-1 Higashi Tsukuba, Ibaraki 305, Japan. E-mail: kitajima@nibh.go.jp. Peter G. Polson, Institute of Cognitive Science, University of Colorado, CB 344, Boulder, CO 80309–0344. E-mail: ppolson@psych.colorado.edu.

HCI Editorial Record. First manuscript received April 5, 1996. Revisions received September 5, 1996, and March 3, 1997. Accepted by Richard Young. Final manuscript received April 28, 1997. — *Editor*

REFERENCES

Anderson, J. R. (1993). *Rules of the mind.* Hillsdale, NJ: Lawrence Erlbaum Associates, Inc.

Bovair, S., & Kieras, D. E. (1985). A guide to propositional analysis for research on technical prose. In B. K. Britton & J. B. Black (Eds.), *Understanding expository text* (pp. 315–362). Hillsdale, NJ: Lawrence Erlbaum Associates, Inc.

Carroll, J. M. (1990). *The Nuremberg funnel: Designing minimalist instruction for practical computer skills.* Cambridge, MA: MIT Press.

Compeau, D., Olfman, L., Sein, M., & Webster, J. (1995). End-user training and learning. *Communications of the ACM, 38,* 24–26.

Cummins, D. D., Kintsch, W., Reusser, K., & Weimer, R. (1988). The role of understanding in solving word problems. *Cognitive Psychology, 20,* 405–438.

Delarosa, D. (1986). A computer simulation of children's arithmetic word problem solving. *Behavior Research Methods, Instruments, and Computers, 18,* 147–154.

Engelbeck, G. E. (1986). *Exceptions to generalizations: Implications for formal models of human–computer interaction.* Unpublished master's thesis, Department of Psychology, University of Colorado, Boulder.

Ernst, G. W., & Newell, A. (1969). *GPS: A case study in generality and problem solving.* New York: Academic.

Fletcher, C. R. (1985). Understanding and solving arithmetic word problems: A computer simulation. *Behavior Research Methods, Instruments, and Computers, 17,* 565–571.

Franzke, M. (1994). *Exploration, acquisition, and retention of skill with display-based systems.* Unpublished doctoral dissertation, Department of Psychology, University of Colorado, Boulder.

Franzke, M. (1995). Turning research into practice: Characteristics of display-based interaction. *Proceedings of the CHI'95 Conference on Human Factors in Computing Systems,* 421–428. New York: ACM.

Franzke, M., & Rieman, J. (1993, September). Natural training wheels: Learning and transfer between two versions of a computer application. *Proceedings of the Vienna Conference on Human–Computer Interaction '93.* Vienna.

Greeno, J. G., & Simon, H. A. (1988). Problem solving and reasoning. In R. C. Atkinson, R. Herrnstein, G. Lindzey, & R. D. Luce (Eds.), *Steven's handbook of experimental psychology* (pp. 589–639). New York: Wiley.

Howes, A. (1994). A model of the acquisition of menu knowledge by exploration. *Proceedings of the CHI'94 Conference on Human Factors in Computing Systems,* 445–451. New York: ACM.

Howes, A., & Young, R. (1996). Learning consistent, interactive, and meaningful task-action mappings: A computational model. *Cognitive Science, 20,* 301–356.

Hudson, T. (1983). Correspondences and numerical differences between disjoint sets. *Child Development, 54,* 84–90.

Hutchins, E. L., Hollan, J. D., & Norman, D. A. (1986). Direct manipulation interfaces. In D. A. Norman & S. W. Draper (Eds.), *User centered system design* (pp. 87–124). Hillsdale, NJ: Lawrence Erlbaum Associates, Inc.

Kintsch, W. (1974). *The representation of meaning in memory.* Hillsdale, NJ: Lawrence Erlbaum Associates, Inc.

Kintsch, W. (1988). The role of knowledge in discourse comprehension: A construction–integration model. *Psychological Review, 95,* 163–182.

Kintsch, W. (in press). *Comprehension: A paradigm for cognition.* Cambridge, England: Cambridge University Press.

Kintsch, W., & Greeno, J. G. (1985). Understanding and solving word arithmetic problems. *Psychological Review, 92,* 109–120.

Kintsch, W., & van Dijk, T. A. (1978). Toward a model of text comprehension and production. *Psychological Review, 85,* 363–394.

Kintsch, W., & Welsch, D. M. (1991). The construction–integration model: A framework for studying memory for text. In W. E. Hockley & S. Lewandowsky (Eds.), *Relating theory and data: Essays on human memory* (pp. 367–385). Hillsdale, NJ: Lawrence Erlbaum Associates, Inc.

Kitajima, M. (1996). *Model-based analysis of required knowledge for successful interaction with a novel display* (ICS Technical Report 96–03). Boulder, CO: Institute of Cognitive Science.

Kitajima, M., & Polson, P. G. (1995). A comprehension-based model of correct performance and errors in skilled, display-based human–computer interaction. *International Journal of Human–Computer Systems, 43,* 65–99.

Kitajima, M., & Polson, P. G. (1996). A comprehension-based model of exploration. *Proceedings of the CHI'96 Conference on Human Factors in Computing Systems,* 324–331. New York: ACM.

Kitajima, M., & Polson, P. G. (1997). LICAI+: A comprehension-based model of learning for display-based human–computer interaction. *Conference companion of Human Factors in Computing Systems CHI'97,* 333–334. New York: ACM.

Mack, R. L., Lewis, C. H., & Carroll, J. M. (1983). Learning to use word-processors: Problems and prospects. *ACM Transactions on Office Information Systems, 1,* 254–271.

Mannes, S. M., & Kintsch, W. (1991). Routine computing tasks: Planning as understanding. *Cognitive Science, 15,* 305–342.

Mross, E. F., & Roberts, J. O. (1992). *The construction–integration model: A program and manual* (ICS Technical Report 92–14). Boulder, CO: Institute of Cognitive Science.

Newell, A. (1990). *Unified theories of cognition.* Cambridge, MA: Harvard University Press.

Norman, D. A. (1986). Cognitive engineering. In D. A. Norman & S. W. Draper (Eds.), *User centered system design* (pp. 31–61). Hillsdale, NJ: Lawrence Erlbaum Associates, Inc.

Payne, S. J., Squibb, H. R., & Howes, A. (1990). The nature of device models: The yoked state hypothesis and some experiments with text editors. *Human–Computer Interaction, 5,* 415–444.

Polson, P. G., & Lewis, C. (1990). Theory-based design for easily learned interfaces. *Human–Computer Interaction, 5,* 191–220.

Polson, P. G., Lewis, C., Rieman, J., & Wharton, C. (1992). Cognitive walkthroughs: A method for theory-based evaluation of user interfaces. *International Journal of Man–Machine Studies, 36,* 741–773.

Raaijmakers, J. G., & Shiffrin, R. M. (1981). Search of associative memory. *Psychological Review, 88,* 93–134.

Rieman, J. (1994). *Learning strategies and exploratory behavior of interactive computer users.* Unpublished PhD dissertation, Department of Computer Science, University of Colorado, Boulder.

Rieman, J. (1996). A field study of exploratory learning strategies. *ACM Transactions on Computer–Human Interaction, 3,* 189–218.

Rieman, J., Lewis, C., Young, R. H., & Polson, P. G. (1994). Why is a raven like a writing desk? Lessons in interface consistency and analogical reasoning from two cognitive architectures. *Proceedings of the CHI'94 Conference on Human Factors in Computing Systems,* 438–444. New York: ACM.

Rieman, J., Young, R. M., & Howes, A. (1996). A dual space model of iteratively deepening exploratory learning. *International Journal of Human–Computer Studies, 44,* 743–775.

Simon, H. A., & Hayes, J. R. (1976). The understanding process: Problem isomorphs. *Cognitive Psychology, 8*(2), 165–190.

Terwilliger, R. B., & Polson, P. G. (1997). Relationships between users' and interfaces' task representations. *Proceedings of the CHI'97 Conference on Human Factors in Computing Systems,* 99–106. New York: ACM.

van Dijk, T. A., & Kintsch, W. (1983). *Strategies of discourse comprehension.* New York: Academic.

Wharton, C., & Kintsch, W. (1991). An overview of the construction–integration model: A theory of comprehension as a foundation for a new cognitive architecture. *SIGART Bulletin, 2,* 169–173.

Wharton, C., Rieman, J., Lewis, C., & Polson, P. G. (1994). The cognitive walkthrough method: A practitioner's guide. In J. Nielsen & R. Mack (Eds.), *Usability inspection methods* (pp. 105–140). New York: Wiley.

Yourdon, E. (1989). *Structured walkthroughs* (4th ed.). Englewood Cliffs, NJ: Author.

HUMAN–COMPUTER INTERACTION, 1997, Volume 12, pp. 391–438
Copyright © 1997, Lawrence Erlbaum Associates, Inc.

An Overview of the EPIC Architecture for Cognition and Performance With Application to Human–Computer Interaction

David E. Kieras and **David E. Meyer**
University of Michigan

ABSTRACT

EPIC (**E**xecutive **P**rocess–**I**nteractive **C**ontrol) is a cognitive architecture especially suited for modeling human multimodal and multiple-task performance. The EPIC architecture includes peripheral sensory-motor processors surrounding a production-rule cognitive processor and is being used to construct precise computational models for a variety of human–computer interaction situations. We briefly describe some of these models to demonstrate how EPIC clarifies basic properties of human performance and provides usefully precise accounts of performance speed.

David E. Kieras is a cognitive psychologist and computer scientist whose speciality is computational modeling of human cognition and performance with an emphasis on application to human–computer interaction and other system design problems; he is an Associate Professor in the Department of Electrical Engineering and Computer Science at the University of Michigan. **David E. Meyer** is a mathematical and experimental cognitive psychology scientist with interests in human cognition and action; he is a Professor in the Cognition and Perception Program of the Department of Psychology at the University of Michigan.

CONTENTS

1. INTRODUCTION

Cognitive Architectures.   A cognitive architecture is a theoretical structure and set of mechanisms for human cognition, within which models for
specific tasks and phenomena can be constructed. Since the proposals of
Anderson (1976) and Laird, Rosenbloom, and Newell (1986), cognitive
architectures have become recognized as the fundamental theoretical ap-

proach in cognitive psychology. An architecture proposal is a synthesis of theoretical concepts that attempts to subsume a variety of specific models and mechanisms into a single coherent whole. When the architecture is represented computationally, its implications and applicability can be easily and rigorously explored and tested. Progress in rigorous cognitive theory requires the development of more comprehensive and accurate computational cognitive architectures.

Cognitive Architecture and Human–Computer Interaction (HCI). The significance of cognitive architectures for the more practical concerns of HCI and user-interface design lies in two areas. First, to the extent that the key phenomena of relevance to HCI can be captured in an architecture, the architecture acts as a codified theoretical summary of the phenomena. Such a codification can then be learned and applied by HCI researchers and practitioners much more easily than the traditional approach of studying and attempting to apply a vast collection of isolated phenomena, individual experimental results, and small-scale models. Second, if the architecture supports constructing models for tasks easily enough and makes accurate enough predictions of task performance, the cognitive architecture then provides a foundation for engineering models for evaluating user-interface designs early in the development process, which can provide valuable usability information in addition to traditional user testing methods. For example, as discussed by John and Kieras (1996), the various extant members of the GOMS family of engineering models are based on some simple cognitive architectures but have been useful in interface design and evaluation. By developing more sophisticated architectures that have predictive power in more complex situations, we should be able to develop more accurate and more comprehensive engineering models to aid in HCI design.

The EPIC (Executive Process–Interactive Control) Architecture. This article provides an overview of the EPIC architecture being developed by Kieras and Meyer for modeling human cognition and performance (Kieras, Wood, & Meyer, 1997; Meyer & Kieras, 1997a, 1997b). EPIC is similar in spirit to the Model Human Processor (MHP; Card, Moran, & Newell, 1983), but EPIC incorporates many recent theoretical and empirical results about human performance in the form of a software framework for computer simulation modeling. Using EPIC, a model can be constructed that represents the general procedures required to perform a complex multimodal task as a set of production rules. When the model is supplied with the external stimuli for a specific task, it will then execute the procedures in whatever way the task requires, thus simulating a human's performing the task and generating the predicted actions in simulated real time. EPIC is an architecture for constructing models of performance. It is not yet a learning system and so has no mechanisms for learning how to perform a task. Rather, the purpose of EPIC is to represent in detail the perceptual, motor, and cognitive constraints on the human ability to perform tasks.

Like most cognitive architectures, EPIC was not developed primarily for addressing HCI problems but is a larger scientific endeavor to represent important theoretical concepts of human intelligence or abilities. However, because HCI is a subset of human performance, a good proposal for a cognitive architecture should allow one to analyze and compare interface designs and then recommend and evaluate improvements. At this time, EPIC is mainly useful as a research system for exploring human performance limitations that determine the effects of a particular interface design, both at low levels of specific interaction techniques and at high levels of systems that support complex task performance in multimodal time-stressed domains. In the future, it should be possible to develop EPIC-based design analysis techniques that can be routinely applied in system design.

Organization of This Article. In this article, we describe the rationale for the development of EPIC, summarize the architecture, and discuss some important general modeling issues. Then, we illustrate EPIC's contributions to HCI with a series of application vignettes—brief examples showing how EPIC can be used in both predictive and explanatory modes to address both elementary aspects of interface design and complex phenomena related to human interaction with semi-automated systems. We use these vignettes because our goal in this article is not to exhaustively explore one application of the EPIC architecture but rather to present and justify the architecture by illustrating its wide applicability.

2. THE EPIC (EXECUTIVE PROCESS–INTERACTIVE CONTROL) ARCHITECTURE

2.1. Fundamental Motivations for Developing EPIC

Embodied Cognition

Historically, the proposals for computational models of human cognition both in cognitive psychology and artificial intelligence have tended to emphasize the purely cognitive aspects of the human system, finessing the details of how the human perceives the environment or acts upon it. Such an approach can be seriously misleading. For example, human visual capacity to detect and recognize objects is not uniform but varies with the distance on the retina from the fovea. The retina can be oriented through a motor system, the oculomotor mechanism, to control what part of the visual environment can be accessed in detail. Thus, in many tasks, the availability of the stimulus depends on the details of when, how, or whether the eye is oriented toward the stimulus. Likewise, conventional cognitive theory has tended to assume that, after the human decides to act, there are no fundamental problems in carrying out the intended action. However, the human motor system is quite complex in its own right and interacts strongly with the cognitive and perceptual systems (Rosenbaum, 1991). For example, different movements can take a substantial amount of time to execute, and this time can depend heavily on the details of the required movements and the history of previous, possibly interfering, movements. Furthermore, response execution can make demands on the perceptual system as well, the most extreme being the need for vision in aimed movements; thus, making a response may interfere with collecting the visual information needed for the next part of the task.

More recently, some of the computational cognitive architectures have begun to be *embodied,* to include some of the constraints imposed by the perceptual-motor system. For example, Newell (1990) explicitly included such constraints in his outline of a cognitive architecture, and some effort has been made to include perceptual-motor mechanisms in more recent work with the Soar architecture (Laird et al., 1986) since its introduction. Likewise, the ACT–R architecture proposed by Anderson (1993) has begun to include some perceptual-motor mechanisms. Our first goal in developing EPIC is not only to incorporate key facts about human perceptual and motor constraints into a theory of human cognition but also to give the perceptual and motor mechanisms equal status with cognition in accounting for human performance. Thus, EPIC's production-rule cognitive processor is surrounded by perceptual and motor processors, whose time course of processing is represented in some detail based on the current human performance research literature. How long it takes an EPIC model to do a task depends intimately on how EPIC's eyes, perceptual mechanisms, and effectors are used in the task. At this level of detail, the interactions between processors during task execution can be remarkably subtle, so the representation of task timing in a computational simulation model is critical to understanding human performance.

Computational Models of Both Performance and Cognition

Our second motivation for developing EPIC is a corollary of the first—to more fully extend the current cognitive architecture computational modeling paradigm to the field of human attention and performance. Although some computationally realized models of human performance have been available for years (e.g., HOS, SAINT; see Elkind, Card, Hochberg, & Huey, 1989; McMillan et al., 1989), these models generally do not have the form of cognitive architectures so much as being analytic tools for practical system design—a form of the engineering models discussed by John and Kieras (1996). The field of human performance itself has suffered from a lack of computational modeling, although historically, it is one of the most extensively researched and most practically useful of psychological fields. Most research on human performance has been conducted at the level of qualitative interpretation of empirical results and verbally expressed and evaluated theory. One symptom of this lack of the detailed rigor available with computational models is that key issues concerning the fundamentals of human information processing have remained unresolved for many years, such as the status of the single-channel-versus-multiple-resource debate discussed by Meyer and Kieras (1997a). Thus, another goal of developing EPIC has been to advance the state of psychological theory in a theoretically underdeveloped area.

An important benefit of working in the human performance domain is that this empirical literature abounds in quantitative data that typically have a precision and detail not found in more purely cognitive task domains. Unlike many traditional cognitive modeling efforts, ours is committed to obtaining detailed and quantitatively accurate fits to empirical data in a variety of performance task domains. One reason for making this effort is that for models of performance to be practically useful in system design problems, they must be reasonably accurate. A more fundamental reason is that trying to match detailed data with quantitative accuracy serves as a powerful constraint in constructing models for phenomena. That is, a key function of a cognitive architecture is to provide some theoretical constraint on the possible models (see Newell, 1990). Trying to match detailed quantitative data acts as a further constraint, further reducing the number of arbitrary decisions to be made in constructing an EPIC model for a task. The constraints imposed by the combination of the detailed quantitative effects in the empirical data, the task structure, and the fixed architecture mean that there are relatively few "degrees of freedom" in constructing a model that fits well (see Meyer & Kieras, 1997b, for further discussion).

The Executive Process and Multiple-Task Performance

A third motivation for developing EPIC has been to explore the mechanism and the role of executive processes, which control and supervise other cognitive processes, analogous to the "supervisor" in a computer operating system (see Meyer & Kieras, 1997a). Theorists of human performance have presented various proposals about the nature of the executive process. Unfortunately, these proposals have usually lacked either a coherent theoretical basis or a computational representation. These theorists have also often proposed that the executive process is implemented via some type of special mechanism that sits outside or above the regular cognitive system and presumably has its own principles of operation.

However, since proper supervision of behavior is simply a form of skill, we have sought to represent the executive process in the same way as other forms of skill—just as the supervisory component of a computer operating system is just another computer program.

A good approach to understanding both the nature of the executive process and the details of the human cognitive architecture is to understand multiple-task performance. In multiple-task situations, the human has to perform two or more tasks simultaneously; the overall task situation can be subdivided into two or more tasks, each of which can be meaningfully performed in isolation (one is not a logical subtask of the other), and the tasks are performed over the same period of time. A good example of a multiple-task situation occurs in an airplane cockpit; a pilot may need to simultaneously pilot the aircraft and track an enemy target on a radar display. The main problem confronting the human is to execute the independent tasks in a coordinated fashion that meets some constraints on overall performance, such as giving one task priority over the other.

The literature on multiple-task performance is extensive and is not summarized here; for a review, see Gopher and Donchin (1986) and Meyer and Kieras (1997a). Of course, human information processing is limited in capacity, and a single-channel bottleneck has traditionally been assumed. Nevertheless, humans can do multiple tasks, sometimes impressively well, and their ability to do so depends strongly on the specific combinations of tasks involved. The *multiple-resource theory* is an attempt to summarize these dependencies, which pose a fundamental theoretical dilemma about how to reconcile the complex patterns of people's multitasking abilities with some notion that the overall capacity of the human system is limited. With EPIC, however, we do not make the assumption that central capacity for cognitive processing is limited. Such an assumption is traditional but lacks both empirical and metatheoretical justification. In contrast, we assume that limitations on human ability are all structural; that is, performance of tasks may be limited by constraints on peripheral perceptual and motor mechanisms or by limited verbal working memory capacity, rather than by a pervasive limit on cognitive-processing capacity. The executive strategy has the responsibility of meeting the performance requirements of the tasks in spite of these structural limitations.

To meet performance goals, the executive process must coordinate the use of the perceptual, cognitive, and motor resources of the system so that the tasks can be conducted with the proper relative priority and speed. Multiple-task situations stress human capabilities very seriously, and so the observed patterns of behavior provide clear insights into the abilities and limitations of the human information-processing system architecture. Yet, despite the practical importance of multiple-task performance, the empirical and theoretical understanding of multiple-task performance has been quite limited. Nevertheless, EPIC models have been successful in accounting with unprecedented accuracy for performance in laboratory versions of multiple-task situations (Kieras & Meyer, 1995; Meyer & Kieras, 1997a, 1997b). In addition, understanding the allocation of resources in multiple-task situations contributes to understanding single-task situations; to maximize performance, the executive process must allocate processing resources to different parts of the single task in a properly coordinated fashion.

2.2. Description of the EPIC Architecture

Overview

Figure 1 shows the overall structure of processors and memories in the EPIC architecture. At this level, although EPIC bears a superficial resemblance to earlier frameworks for human information processing, it incorporates a new synthesis of theoretical concepts and empirical results and so is more comprehensive and more detailed than earlier proposals for human performance modeling (e.g. MHP, HOS, SAINT; see McMillan et al., 1989). EPIC is designed to explicitly couple detailed mechanisms for basic information processing and perceptual-motor activity with a cognitive analysis of procedural skill—namely, that represented by production-system models such as CCT (Bovair, Kieras, & Polson, 1990), ACT-R (Anderson, 1993), and Soar (Laird et al., 1986). Thus, EPIC has a production-rule cognitive processor surrounded by perceptual-motor peripherals; applying EPIC to a task situation requires specifying both the production-rule programming for the cognitive processor and the relevant perceptual and motor-processing parameters. EPIC computational task models are *generative,* in that the production rules supply general procedural knowledge of the task, and, when EPIC interacts with a simulated task environment, the EPIC model generates the specific sequence of serial and parallel human actions required to perform the specific tasks. Rather than reflecting specific task scenarios, the task analysis reflected in the model is general to a class of tasks.

The software for constructing EPIC models is currently implemented in Common LISP, with models typically developed in Macintosh Common Lisp, and then simulation production runs are executed under Franz Allegro Common LISP on a fast Unix workstation. All of the models described or cited in this article have actually been implemented and run to produce the claimed predicted results. Although the simulation software is available to interested researchers, as is a detailed technical description of the architecture,[1] we have focused exclusively on developing the architecture and modeling important tasks that stress the scientific accuracy of the architecture. Thus, at this time, EPIC is not packaged in a

---

1. Available via anonymous ftp at ftp://ftp.eecs.umich.edu/people/kieras/EPICarch.ps.

Figure 1. Overall structure of the EPIC architecture simulation system. Task performance is simulated by having the EPIC model for a simulated human (on the right)
interact with a simulated task environment (on the left) via a simulated interface
between sensory and motor organs and interaction devices. Information flow paths
are solid lines; mechanical control or connections are dashed lines. The processors
run independently and in parallel both with each other and with the task environment module.

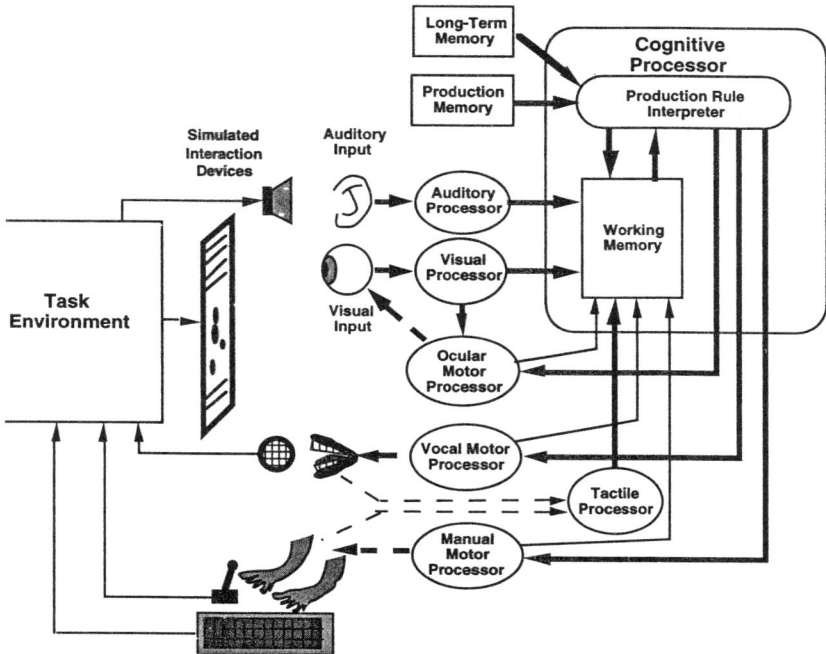

"user-friendly" manner; full-fledged LISP programming expertise is required to use the simulation package, and there is no introductory tutorial
or user's manual.

The EPIC software framework includes not only the modules for simulating a human but also facilities for simulating the interaction of the
human with an external system such as a computer. Figure 1 shows a
simulated task environment (on the left) and a simulated human as described by the EPIC architecture (on the right), with objects such as
simulated screen items and simulated keys making up the physical interface between them. The task environment module assigns physical locations to the interface objects and generates simulated visual events and
sounds that the computer or other entities in the environment produce in
response to the simulated human's behavior. Having a separate environment simulation module greatly simplifies the programming of a complete
simulation and helps enforce the generality of the procedural knowledge

represented in the EPIC model. That is, the task environment module is driven by a task instance description that consists only of the sequence and timing of events external to the human user, and the simulated user must deal with whatever happens in the simulated task environment.

With regard to the EPIC architecture itself (as shown in Figure 1), there is a conventional flow of information from sense organs, through perceptual processors, to a cognitive processor (consisting of a production rule interpreter and a working memory), and finally to motor processors that control effector organs. EPIC goes beyond MHP by specifying separate perceptual processors with distinct processing-time characteristics for each sensory modality and separate motor processors for vocal, manual, and oculomotor (eye) movements. There are feedback pathways from the motor processors, as well as tactile feedback from the effectors, which are important in coordinating multiple tasks. The declarative–procedural knowledge distinction of the "ACT-class" cognitive architectures (e.g., Anderson, 1976) is represented in the form of separate permanent memories for production rules and declarative information. Working memory (WM) contains all of the temporary information needed for and manipulated by the production rules, including control items such as task goals and sequencing indices, and also conventional WM items, such as representations of sensory inputs. WM has separate partitions for different types of information, such as auditory WM, visual WM, the *control store,* and so forth. The structure of WM and the properties of the processors are described in more detail later. When numeric values are given for various time parameters, they are labeled as either *standard values* that we assume are fixed in all applications of the architecture or *typical values* that may vary depending on the properties of a specific task situation. Standard values are based on our reading of the human performance literature; although they may be revised and refined, they are supposed to hold across all applications of the architecture. The nonstandard parameter values need to be estimated to model a specific task, but we hope that, with additional modeling experience and focused empirical studies, collections of typical parameter values will become available for use in constructing new models.

Perceptual Processors

A single stimulus input to a perceptual processor can produce multiple outputs to be deposited in WM at different times. The perceptual processors in EPIC are simple "pipelines," in that an input produces an output at a certain later time, with no "moving window" time-integration effect as assumed by MHP. The tactile perceptual processor handles movement feedback from effector organs; this feedback can be important in coordinating multiple tasks (Meyer & Kieras, 1997a, 1997b) but is not elaborated further here.

Visual Processor.   EPIC's model of the eye includes a retina that determines what kind of sensory information is available about visual objects in the environment based on the distance (in visual angle) on the retina between the object and center of the fovea. EPIC's current highly simplified model of the retina contains three zones, each with a standard radius: the fovea (1°), the parafovea (10°), and the periphery (60°). Certain information (e.g., contents of character strings) might be available only in the fovea, whereas cruder information (e.g., whether an area of the screen is filled with characters) is available in the parafovea. Only severely limited information (e.g., location of objects; whether an object has just appeared) is available in peripheral vision. Of course, the exact availability of visual information in different areas of the retina depends on the specific physical properties of the stimulus. For example, a large isolated character might be discriminable many degrees away from the fovea, while reading words embedded in text displayed in small type would require that the words be in or very close to the fovea. Unfortunately, the human performance literature does not appear to contain a body of well-parameterized results on the properties of nonfoveal vision, meaning that the exact time and availability parameters of visual stimuli must be estimated for new task-specific models.

In EPIC's visual working memory, the visual perceptual processor maintains a representation of which objects are visible and what their properties are. Visual working memory is "slaved" to the visual situation; it is kept up-to-date as objects appear, disappear, change color, and so forth or as eye movements or object movements change what visual properties are available from the retina. In response to visual events, the visual processor can produce multiple outputs with different timings. When an object appears, the first output is a representation that a perceptual event has been detected (standard delay = 50 msec), followed later by a representation of sensory properties (e.g., shape; standard delay = 100 msec) and still later by the results of pattern recognition, which might be task-specific (e.g., a particular shape represents a left-pointing arrow; typical delay = 250 msec).

Auditory Processor.   The auditory perceptual processor accepts auditory input and then outputs to working memory representations of auditory events (e.g., speech) that disappear after a time. For example, a short tone signal produces, first, an item corresponding to the onset of the tone (standard delay = 50 msec); then, later, an item corresponding to a discriminated frequency of the tone (typical delay = 250 msec); and, last, an offset item (standard delay = 50 msec). For simplicity, such items simply disappear from memory after a fixed time (typical delay = 4 sec).

Speech input is represented as items for single words in auditory working memory. The auditory perceptual processor requires a certain time to recognize input words (typical delay = 150 msec after acoustic stimuli are present) and produces representations of them in auditory working memory. These items then disappear after a time, the same as other auditory input. To represent the sequential order of the speech input, the items contain arbitrary symbolic tags for previous item and next item that link the items in sequence. Thus, a speech input word carries a certain next-tag value, and the next word in the sequence is the item that contains the same tag value as its previous tag. Using these tags, a set of production rules can step through the auditory working memory items for a series of spoken words, processing them one at a time. For example, one of the models described in this article processes a spoken telephone billing number by retrieving the recognized code for each digit in the tag-chained sequence and using it to specify a key press action. Available empirical literature on auditory perception lacks comprehensive results, so many auditory perceptual parameters must be estimated during task-specific model construction.

Cognitive Processor

Production Rules and Cycle Time.   The cognitive processor is programmed in terms of production rules, and so an EPIC model for a task must include a set of production rules that specify what actions in what situations are performed to do the task. Example production rules for the models described in this article are presented later. EPIC uses the parsimonious production system (PPS) interpreter, which is especially suited to task modeling work (Bovair et al., 1990). PPS rules have the format (<rule-name> IF <condition> THEN <actions>). The rule condition can test only the contents of the production-system working memory. The rule actions can add and remove items from working memory or send a command to a motor processor.

The cognitive processor operates cyclically. At the beginning of each cycle, the contents of working memory are updated with the output from perceptual processors and the previous cycle's modifications; at the end of each cycle, the contents of the production-system working memory are updated, and commands are sent to the motor processors. The mean duration of a cycle is a standard 50 msec. The cognitive-processor cycles are not synchronized with external stimulus-and-response events. Inputs from the perceptual processors are accessed only intermittently, when the production-system working memory is updated at the start of each cycle. Any input that arrives during the course of a cycle must therefore wait temporarily for service until the next cycle begins. This is consistent with a variety of phenomena, such as the apparent temporal granularity of perceived stimulus successiveness (Kristofferson, 1967). EPIC also can run in a mode in which the cycle duration is stochastic, with a standard mean value of 50 msec and all other time parameters scaled to this stochastic value; the variance of the stochastic distribution of cycle time is chosen to produce a coefficient of variation of about 20% for a simple reaction time, corresponding to the typical observed value.

Cognitive Parallelism. Most traditional production-system architectures require that only one production rule can be fired at a time and that only its actions will be executed. Should more than one rule have matching conditions, some kind of conflict-resolution mechanism is required to choose which rule to fire. Soar (Laird et al., 1986) is perhaps the most complex, in that the production rules only propose operators to apply, and so many rules can be fired at once, and then a separate process decides which single candidate operator to apply. However, PPS has a radical and very simple policy: On each cognitive-processor cycle, PPS will fire all rules whose conditions match the contents of working memory and will execute all of their actions. Thus, EPIC models may have true parallel cognitive processing at the production-rule level; multiple "threads" or processes can be represented simply as sets of rules that happen to run simultaneously.

The multiprocessing ability of the cognitive processor, together with the parallel operation of all the perceptual-motor processors, means that EPIC models for multiple-task performance do not incorporate a gratuitous assumption of limited central-processing capacity or of a central-processing "bottleneck." It is critical to be clear on exactly what is or is not being claimed. Although EPIC has no built-in limit on how many strategies or processes the cognitive processor can be executing simultaneously, the rate of execution is limited, the perceptual-motor mechanisms are of course limited, and any memory mechanisms involved in the task are limited. For example, the reason why a person cannot perform two long divisions in his or her head simultaneously is that a limited structural resource is involved—namely, verbal short-term memory. For example, the eyes can fixate on only one place at a time, and the two hands are bottlenecked through a single processor. Thus, a task that demands manual responses to visual stimuli distributed over a wide space will result in severely limited human performance, even if the purely cognitive demands are trivial. In contrast, people can, and often do, perform multiple cognitive tasks simultaneously, such as collecting one's slides while answering questions at the end of a talk, as long as the strategies are otherwise compatible.

Omitting a central-capacity limit or bottleneck might seem to be a radical recasting of conventional cognitive theory, but this claim is actually consistent with a long-standing line of empirical and theoretical discussion in the human performance field that challenges the traditional assumption of limited central capacity (see Meyer & Kieras, 1997a). Our own detailed quantitative modeling of a variety of multiple-task data (Meyer & Kieras, 1997b) shows that EPIC's assumptions about the nature of cognitive and perceptual-motor limitations are quite consistent with a large variety of empirical data.

This decision about the nature of human limitations is also a matter of scientific tactics: Our theoretical strategy has been to make some radical simplifying assumptions and then explore their consequences through modeling, complicating the architecture only as required. Thus, we start with the known limitations of human memory and perceptual-motor mechanisms and adopt less apparent limitations only if the data compel us. Thus far, our simple and radical set of assumptions about the nature of multiple-task processing limitations has held up well.

Working Memory. EPIC's production-system working memory is in effect partitioned into several working memories.[2] Visual, auditory, and tactile memory contain the current information produced by the corresponding perceptual processors. The timing and duration of these forms of working memory are described earlier. Motor working memory contains information about the current state of the motor processors, such as whether a hand movement is in progress. This information is updated on every cycle.

Two other forms of working memory deserve special note. These forms are amodal, in that they contain information not directly derived from sensory or motor mechanisms. One amodal working memory is the *control store,* which contains items that represent the current goals and the current steps within the procedures for accomplishing the goals. An important feature of PPS is that control information is simply another type of working memory item and so can be manipulated by rule actions; this is critical for modeling multiple-task performance, in that production rules for an executive process can control subprocesses by manipulating the control store.

The second amodal working memory, simply termed "general WM," can be used to store miscellaneous task information. At this time, EPIC does not include assumptions about the decay, capacity, and representational properties of general working memory. Our research strategy in developing EPIC has been to see what constraints on the nature of this general WM are required to model task performance in detail rather than to follow the customary strategy in cognitive modeling of assuming these

---

2. EPIC's working memory structure is not "hard-wired" into PPS. PPS actually has only a single working memory, which could more clearly be termed the *database* for the production rules. PPS can be used as a multiple-memory system simply by following a convention such as the first term in a database item indicating the "type" of memory item, as in the examples later in this article. Likewise, the format of items in working memory or the required contents of rule conditions are not fixed in the architecture; we have preferred to develop such restrictions through modeling experience rather than prematurely prescribe them in the architecture.

constraints in advance. Such capacity and loss assumptions for these memory systems do not seem to be required to account for the time course of performance in tasks modeled in EPIC thus far; other limitations determined by the perceptual and motor systems appear to dominate performance. These latter substantial but underappreciated limitations would have been obscured by gratuitous assumptions about central-processing capacity or working memory (see Meyer & Kieras, 1997a, 1997b, for more discussion). For similar reasons, at this time EPIC assumes that information is not lost from the control store, and there is no limit on the capacity of the control store. Research is underway to explore how loss or corruption of information in EPIC's working memories might account for the occurrence and properties of human errors during task performance.

Motor Processors

The EPIC motor processors are much more elaborate than those in the MHP, producing a variety of simulated movements of different effector organs and taking varying amounts of time to do so. As shown in Figure 1, there are separate processors for the hands, eyes, and vocal organs, and all can be active simultaneously. The cognitive processor sends a command to a motor processor that consists of a symbolic name for the type of desired movement and any relevant parameters, and the motor processor then produces a simulated movement with the proper time characteristics. The various processors have similar structures but different timing properties and capabilities based on the current human performance literature in motor control (Rosenbaum, 1991). The manual motor processor has many movement forms, or styles, but the two hands are bottlenecked through a single manual processor, and so normally can be operated either one at a time or synchronized with each other. The oculomotor processor generates eye movements either upon cognitive command or in response to certain visual events. The vocal motor processor produces a sequence of simulated speech sounds given a symbol for the desired utterance.

Movement Preparation and Execution.   The various motor processors represent movements and movement generation in the same basic way. Current research on movement control (Rosenbaum, 1980, 1991) suggests that movements are specified in terms of movement features, and the time to produce a movement depends on its feature structure as well as its mechanical properties.

The overall time to complete a movement can be divided into a *preparation* phase and an *execution* phase. The preparation phase begins when the motor processor receives the command from the cognitive processor. The motor processor recodes the name of the commanded movement into a set of movement features, whose values depend on the style and characteristics of the movement, and then generates the features, taking a standard 50 msec for each one. The time to generate the features depends on how many features can be reused from the previous movements (repeated movements can be initiated sooner) and how many features have been generated in advance. After the features are prepared, the execution phase begins with an additional standard delay of 50 msec to initiate the movement followed by the actual physical movement. The time to physically execute the movement depends on its mechanical properties both in terms of which effector organ is involved (e.g., eye vs. hand) and type of movement to be made (e.g., one-finger flexion to press a button under the finger vs. a pointing motion with a mouse).

The movement features remain in the motor processor's memory, so future movements that share the same features can be performed more rapidly. However, there are limits on whether features can be reused; for example, if a new movement is different in style from the previous movement, all of its features must be generated anew. Also, if the task permits the movement to be anticipated, the cognitive processor can command the motor processor to prepare the movement in advance by generating all of the required features and saving them in motor memory. Then, when it is time to make the movement, only the initiation time is required to commence the mechanical execution of the movement.

Finally, a motor processor can prepare the features for only one movement at a time and will reject any subsequent commands received during the preparation phase, but the preparation for a new movement can be done in parallel with the physical execution of a previously commanded movement. Once prepared, the movement features are saved in motor memory until the previous execution is complete, and the new movement is then initiated. The cognitive-processor production rules can exploit this capability by sending a motor processor a new movement command as soon as it is ready to begin preparing the features for the new movement. The result can be a series of very rapid movements whose total time is little more than the sum of their initiation and mechanical execution times.

An Example of Motor-Processor Operation.   To strike a key using a one-finger peck movement style (like that used in "hunt-and-peck" typing), the cognitive processor commands the manual motor processor to perform a peck movement with a finger (e.g., the right index) to a specified object in the physical environment (the key). This movement style involves five features: peck style, hand, finger, direction, and extent of motion, which is the distance between the current location of the designated finger and the location of the target object. If a previous movement was also a peck movement with the same hand and finger, only the direction and extent might have to be generated anew. If the movement is also similar in direction and extent to the previous movement, then all of the features could be reused; none would have to be generated anew. After the features are generated, the movement is initiated. The time required to physically execute the movement to the target is given by Welford's form of Fitts' law (see Card et al., 1983, chap. 2), with a standard minimum execution time of 100 msec, reflecting that, for small movements to large targets, there is a physiologically determined lower bound on the actual duration of a muscular movement. After the simulated finger hits the key, it is left in the location above the key to await the next movement.

Manual Motor Processor. EPIC's manual motor processor represents several movement styles, including punching individual keys or buttons already known to be below the finger, pecking keys that may require some horizontal motion, posing the entire hand at a specified location, pattering two-finger movements one after the other, poking at an object (e.g., on a touch screen), pointing at an object with a mouse, and plying a control (e.g., a joystick) to position a cursor onto an object. Each style of movement has a particular feature structure and an execution-time function that specifies how long the mechanical movement takes to actuate the device in the task environment.

Vocal Motor Processor. EPIC's vocal motor processor is not very elaborated at this time; it is based on the minimal facilities needed to model certain dual-task situations (see Meyer & Kieras, 1997a, 1997b). A more complete version of the vocal motor processor would be able to produce extended utterances of variable content, taking into account that the sequential nature of speech means that movements can be prepared on the fly during the ongoing speech. The current version of EPIC assumes that simple fixed utterances can be designated with a single symbol and require only the preparation of two features before execution begins. The actual production of the sound is assumed to be delayed by about 100 msec after initiation and continues for a time determined by the number of syllables in the words. Further development of the vocal motor processor is planned for the future.

Oculomotor Processor. EPIC's eye movements are produced in two modes, voluntary and involuntary (reflexive). The cognitive processor commands voluntary eye movements, which are saccades to a designated object. A saccade requires generation of up to two features, the direction and extent of the movement from the current eye position to the target object. Execution of the saccade currently is estimated to require a standard 4 msec per degree of visual angle. The oculomotor processor also makes involuntary eye movements, either saccades or small smooth adjustments in response to the visual situation (hence the arrow between the visual perceptual processor and the oculomotor processor in Figure 1). A sudden onset (appearance) of an object can trigger an involuntary saccade. The fovea being somewhat off-center on an object will produce a "centering" movement, which will automatically help zero in on an object after a saccade. Also, the eye will automatically follow a slowly moving object using smooth movements and occasional small saccades (cf. Hallett, 1986). In some of the tasks presented later in this article, EPIC can follow moving objects with a mixture of voluntary and involuntary eye movements. The partial autonomy of the oculomotor processor permits the cognitive processor to choose an object to examine, command that the eye be moved to it, and then leave the details of keeping it centered on the fovea to the oculomotor processor.

2.3. Constructing Models in EPIC

Constraints on Model Construction

Fixed Architecture, Variable Strategies.   The presentation of any modeling approach should document what aspects or parameters of the modeling framework are fixed and are thus supposed to generalize across applications and what aspects or parameters have to be adjusted to fit the data or estimated from data specific to the situation being modeled. In EPIC, the most important fixed aspect is the connections and mechanisms of the EPIC processors, which are supposed to apply without modification across task domains. Our models are thus built by "programming" a fixed and comprehensive architecture with a task strategy expressed in production rules executed by the cognitive processor. We always attempt to explain phenomena in terms of task-specific cognitive strategies before changing the architecture itself. Thus, the key aspect of EPIC that is free to vary in a task-specific way is the task-specific production-rule programming, which is constrained to some extent because it must be written to execute the task correctly and reasonably efficiently.

Fixed and Free Parameters.   The fixed parameters are most time parameters in the processors and the feature structure of the motor processors for individual styles of movement. The model properties and parameters that are then free to vary from model to model or task to task are, first, the task-specific sensory availabilities and perceptual encoding types and times involved in the task (constrained to be similar and constant over similar perceptual events) and second, the styles of movement used to control the device (e.g., touch-typing vs. visually guided pecking), if they are not constrained by the task.

Model Inputs and Outputs.   Similarly, any modeling approach should document what information the model builder has to supply in order to construct the model and what information the constructed model, will then produce in return for the supplied information. To construct an EPIC model, the model builder has to supply the information corresponding to the three free parameters just described—namely:

1. A production-rule representation of the task procedures.
2. Task-specific sensory availabilities and perceptual-processor encodings and timings.
3. Any movement styles not determined by the task requirements.

In addition, the model builder must supply:

4. The simulated task environment, which includes the physical locations and characteristics of relevant objects external to the human.
5. A set of task instances whose execution time is of interest; these instances must specify only environmental events and their timing and are used to control only the environment module of the simulation.

In return for these inputs, the EPIC model will interact with the simulated task environment, generating the predicted sequences of simulated human actions required by each task instance and the predicted time of occurrence of each action. If the production rules were written to describe general procedural knowledge of how to perform the task, these predictions can be generated for any task instance subsumed by these general procedures.

Modeling Multiple Tasks and Executive Processes

The Executive Process.   Some theorists of multiple-task performance postulate an executive-control process that coordinates the separate multiple tasks (e.g., Norman & Shallice, 1986). We do likewise, but a key feature of our approach is that the executive-control process is just another set of production rules. These rules can control other task processes by manipulating information in the control-store partition of the production-system working memory. For example, we assume that each task is represented by a set of production rules that have the task goal appearing in their conditions, and so an executive-process rule can suspend a task by removing its governing goal from the control store and then cause it to resume operation by reinserting the goal. Also, the executive process can cause a task to follow a different strategy by placing in general WM an item for which task rules test, thus enabling one set of rules and disabling another. In addition, the executive process may control sensory and motor peripherals directly (e.g., moving the eye fixation from one point to another) in order to allocate these resources between two tasks. Thus, rather than postulating an executive control mechanism that is somehow different in kind than other cognitive mechanisms, EPIC has a uniform mechanism for the control of behavior both at the executive level and at the detailed level of individual task actions. As a corollary, learning how to coordinate multiple tasks is simply learning another (possibly difficult) skill, as has been proposed by some recent investigators (e.g., Gopher, 1993).

Modeling Elementary Multiple Tasks.   Our first work with EPIC focused on the simplest and most heavily studied dual-task situation in the research literature, the so-called psychological refractory period (PRP) procedure. The PRP procedure consists of two temporally overlapping choice reaction-time tasks; the subject is instructed to make the response for the first task before making the response for the second task. The primary measure of interest is the reaction time (RT) for the second task, which may be affected by the temporal spacing between the two task stimuli. The basic empirical result is that the second response is substantially delayed as the spacing between the two stimuli decreases. The conventional interpretation of this effect (the PRP effect) is that the human has a central response-selection bottleneck, and so the second response cannot be selected or initiated until the first response has been made. However, the details of the effect, and how it depends on other factors such as the stimulus and response modalities of the two tasks, form a complex pattern that has never been satisfactorily explained in any detail.

Meyer and Kieras (1997a, 1997b) provided an exhaustive treatment of the PRP effect using EPIC simulations, and mathematical analyses based on them, to account quantitatively for the results in many published and new experiments. This account interprets the PRP effect as a product of task strategy rather than as a "hard-wired" central bottleneck. In order to conform to the task instructions, subjects must adopt a strategy that postpones initiating the second response until they can ensure that it does not occur before the first response; the magnitude of the delay in the second response depends on how much of the second-task processing can be overlapped with the first task, which in turn depends on the details of the task structure (e.g., whether eye movements are required), the task difficulty, and the task modalities. The EPIC architecture captures the relevant constraints very well; Meyer and Kieras were able to construct models that accounted for the specific patterns of effects in quantitative detail and that revealed the underlying structure of the phenomena. Details and a discussion of recent experiments claiming to refute the EPIC account of PRP are available in Meyer and Kieras (1997a, 1997b).

Lessons From Multiple-Task Modeling.   An immediate insight from the application of the EPIC architecture to multiple-task domains is that there are many possibilities for performing task activities in parallel. That is, in dual-task models using EPIC, the role of the cognitive strategies in coordinating activity between the two tasks is critical to accounting for the observed effects, and, in many situations, these strategies are surprisingly subtle and efficient. In dual-task experiments, the subject is supposed to complete each of two tasks as rapidly as possible, but the higher priority task must be completed before the lower priority task, regardless of the relative speed of the perceptual or motor processing involved in the two tasks. If two tasks require the same motor processor, both perceptual and cognitive processing on the lower priority task can go on while the higher priority task is allowed to control the motor processor. After the motor processor has commenced execution of a higher priority response, the lower priority task can be given control of the motor processor, thereby honoring the task coordination requirements while maximizing speed. If the two tasks involve different motor modalities, portions of the lower priority response can be prepared in advance, so that this response can be made more quickly when its turn comes. If the two tasks compete for the use of both the eyes and the hands, the executive rules can dynamically switch control of the two processors between the two tasks so that their processing is interleaved, minimizing idle time. Thus, in a dual-task situation, the cognitive-processor strategies are responsible for allocating the eyes and the motor processors to the two tasks as needed to maximize overall performance. Similarly, in a single multimodal task with a requirement for speed, the task strategy is responsible for ensuring that the individual processors do their work as soon as possible so as to minimize the total time required for the task.

The Need for Modeling Policies

Model Fitting Versus Performance Prediction.   There are multiple possibilities for how activities in a multimodal task can be overlapped under the EPIC architecture. One way to identify the specific strategy that governs overlapping in a task is to propose a strategy, generate predicted performance under that strategy, compare the predicted performance to empirical data, and repeat until the predicted data match the empirical results. In typical scientific cognitive modeling work devoted to verifying a cognitive architecture and understanding how a task might be done, it is acceptable to arrive at task strategies in this post hoc model-fitting mode. However, after scientific work on cognitive architectures has progressed beyond simple demonstrations of feasibility, success at a priori prediction is required to fully establish the architecture on scientific grounds and to use the architecture in practical settings to analyze the merits of alternative designs. Predicting performance on an a priori basis requires not only a usefully accurate cognitive architecture but also a set of modeling policies for how to choose and represent task strategies on an a priori basis. Developing such policies can only be done by systematically characterizing the space of possible models and testing their accuracy; an example was reported by Kieras et al. (1997), who modeled a task previously studied by Gray, John, and Atwood (1993).

An Example of Performance Prediction via Modeling Policies. In Kieras et al. (1997), EPIC was used to predict performance in a well-practiced multimodal task: Telephone operators collect billing numbers spoken by customers, enter the numbers into a computer workstation, and verify the numbers before allowing the call to proceed. The volume of this work is such that saving a few seconds of work time per call is worth millions of dollars annually in labor costs. Human operators normally overlap speaking and listening to the customer with striking keys and watching for information to appear on the screen. The time taken to handle the call is not simply the sum of the individual activity times but is a complex function of which activities can be overlapped and to what extent. Our EPIC models were constructed on an a priori basis following several modeling policies that start with a simple procedural task analysis which is then translated in a standardized format into a set of EPIC production rules. The predicted task execution times were accurate within 10% to 14%, which is useful in an engineering context. The EPIC architecture accurately represents the perceptual and motor constraints in the task, making it possible to easily construct a model on an a priori basis that predicts the task time accurately enough to aid in choosing between alternative designs. The effort required to construct the EPIC models is fairly modest. In return, the resulting EPIC models can generate predicted execution times for all possible task instances within the scope of the model. Thus, EPIC models appear to be efficient engineering models for multimodal high-performance tasks.

3. EXAMPLES OF APPLYING EPIC TO HUMAN–COMPUTER INTERACTION

Our goal in this article is to present the EPIC architecture and show how it can be applied to a variety of problems in HCI. Thus, in this section, we emphasize the variety by presenting a series of vignettes illustrating current applications of EPIC to various HCI situations, ranging from low-level interaction phenomena to complex interactions with visual displays in dual-task settings. Two recurring themes in this work are the critical role of visual layout and eye movements and the importance of parallel processing or multitask execution, even in simple situations.

3.1. Selecting Items From Menus

Choosing items from a pull-down menu with a mouse is a standard feature of many current interfaces. However, the research and practical design literature does not contain a comprehensive or even very explicit model of how such menu access is done. At first glance, it would seem that the user must first visually scan the list of menu items, looking at each one in turn, and when the desired item is found, make a movement with the mouse to position the cursor there. Other authors (e.g., Sears & Shneiderman, 1994) have made other proposals for menu search that are rather more elaborate but without detailed empirical support, and there is even some support for the notion that menu search can be completely random (Card, 1984), although other results appear to refute it (Lee & MacGregor, 1985). EPIC can be used to represent different hypotheses about how users select menu items, and the predicted results can then be compared to data with high precision. Working with us, Anthony Hornof has begun to model menu access as part of a larger program of research on visual search and visual layout. Hornof's first results suggest that EPIC can be used to address a variety of visual issues in interface design (see also Hornof & Kieras, 1997).

Hornof has modeled performance in a task that was one condition of an experiment by Nilsen (1991) in which subjects selected items from a pull-down menu. In Nilsen's experiment, the subject was shown a digit, and then he or she clicked on a target. This would cause a vertical menu of the digits 1 to 9 to appear in a random order below the cursor position. The subject then pointed to and clicked on the previously designated digit in the menu. The time to select a digit as a function of its location in the randomly ordered menu was fairly linear, with a slope of about 100 msec per item; similar results have been obtained elsewhere.

Serial-Search Model

Hornof constructed a serial-search model for menu-item selection that corresponds to the one-at-a-time hypothesis of visual search. The eye is moved to the next object down the menu, and if that object matches the sought-for item, a pointing movement is initiated to the item; otherwise, the eye is moved to the next object. Figure 2 shows the key production rules in this model. First, the rule IF-NOT-TARGET-THEN-SACCADE-ONE-ITEM waits for the current visual object (bound to the variable ?OBJECT) to have a text LABEL property in visual working memory and fires if this label does not match the sought-for label bound to the variable ?NT. If this rule fires, the DELDB action (delete from the production-system database) and the ADDDB action (add to the database) update the production-system database to make the object below the current item be the new current item, and the SEND-TO-MOTOR action instructs the ocular motor processor to move the eye to this object.

Figure 2. Production rules from the serial search model of menu selection. The first rule repeatedly moves the eye down the menu as long as the text label for the menu item fails to match the sought-for item in working memory. If a matching item is found, the second rule points the mouse cursor to it.

```
(IF-NOT-TARGET-THEN-SACCADE-ONE-ITEM
IF
((GOAL DO MENU TASK)
  (STEP VISUAL-SEARCH)
  (WM CURRENT-ITEM IS ?OBJECT)
  (VISUAL ?OBJECT IS-ABOVE ?NEXT-OBJECT)
  (NOT (VISUAL ?OBJECT IS-ABOVE NOTHING))
  (MOTOR OCULAR PROCESSOR FREE)
  (VISUAL ?OBJECT LABEL ?NT)
  (NOT (WM TARGET-TEXT IS ?NT)))
THEN
((DELDB (WM CURRENT-ITEM IS ?OBJECT))
  (ADDDB (WM CURRENT-ITEM IS ?NEXT-OBJECT))
  (SEND-TO-MOTOR OCULAR MOVE ?NEXT-OBJECT)))

(TARGET-IS-LOCATED-BEGIN-MOVING-MOUSE
IF
((GOAL DO MENU TASK)
  (STEP VISUAL-SEARCH)
  (WM TARGET-TEXT IS ?T)
  (VISUAL ?TARGET-OBJECT LABEL ?T)
  (WM CURSOR IS ?CURSOR-OBJECT)
  (MOTOR MANUAL PROCESSOR FREE))
THEN
((DELDB (STEP VISUAL-SEARCH))
  (ADDDB (STEP MAKE RESPONSE))
  (SEND-TO-MOTOR MANUAL PERFORM
      POINT RIGHT ?CURSOR-OBJECT ?TARGET-OBJECT)))
```

When the text label for this object becomes available, the rule might fire again. However, if the text label for the object matches the sought-for label, the rule TARGET-IS-LOCATED-BEGIN-MOVING-MOUSE will fire next instead. This rule disables both itself and the first rule from firing again by removing the (STEP VISUAL-SEARCH) item from the database, enables the next step in the procedure by adding a STEP item, and sends the manual motor processor an instruction to perform a right-hand movement using the POINT style that positions the mouse cursor onto the target; the execution time follows Fitts' law (see earlier; Card et al., 1983).

Applying this model to Nilsen's (1991) task requires estimating a parameter for how long it takes to recognize the text label for the digits and where on the retina this recognition could be done. Hornof assumed that the digit recognition could be done only in the fovea and that it required 200 msec per digit—a value that had been used for similar stimuli in models for other

Figure 3. Observed and predicted menu selection times. Observed times (solid points and lines) are from Nilsen (1991); predicted times (open points and dotted lines) are explained in the text.

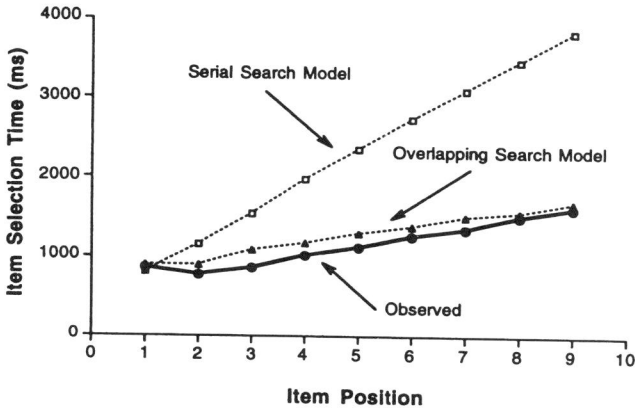

domains. EPIC was run in a simulated version of Nilsen's experiment, and predicted menu selection times were obtained. As shown in Figure 3, the serial-search model seriously misfits the data by predicting a slope of about 380 msec per item, far larger than the observed value of about 107 msec. The reason for the discrepancy is that, according to the EPIC architecture, the serial-search model will require an estimated time of at least 200 msec for the perceptual process to identify the menu item, about 50 msec for a cognitive-processor production rule to fire to initiate the eye movement to the next item, and due to the savings from repeated similar movements, only slightly more than about 50 msec for each subsequent eye movement. Even if the perceptual processing takes much less time (e.g., only 100 msec), the oculomotor and cognitive times are still too long to produce a slope as shallow as 100 msec. In summary, data such as Nilsen's are extremely difficult to reconcile with a model that assumes a strategy of sequentially fixating and deciding about each menu item separately.

Overlapping-Search Model

Hornof next developed a more sophisticated strategy that more fully exploits the parallelism possible in EPIC. A scanning process moves the eye from one item to the next, relying on the parallel perceptual processing "pipeline" to complete the recognition processing of each item. Meanwhile, a separate matching rule waits for the sought-for item to be recognized and appear in working memory; then, it stops the scan and initiates the mouse movement. The relevant rules are shown in Figure 4. The production rule SACCADE-ONE-ITEM moves the eye from item to item as rapidly as possible. Because the movements are repeated, the oculomotor

Figure 4. Production rules from the overlapping-search model. The first rule moves the eye rapidly down the menu, regardless of whether a matching item is present. The second rule halts the scan if a matching item appears in visual working memory and enables the third rule, which moves the eye back to the matching item and launches the mouse-pointing movement to it.

```
(SACCADE-ONE-ITEM
IF
((GOAL DO MENU TASK)
   (STEP VISUAL-SWEEP)
   (WM CURRENT-ITEM IS ?OBJECT)
   (VISUAL ?OBJECT IS-ABOVE ?NEXT-OBJECT)
   (NOT (VISUAL ?OBJECT IS-ABOVE NOTHING))
   (MOTOR OCULAR PROCESSOR FREE)
   )
THEN
((DELDB (WM CURRENT-ITEM IS ?OBJECT))
   (ADDDB (WM CURRENT-ITEM IS ?NEXT-OBJECT))
   (SEND-TO-MOTOR OCULAR MOVE ?NEXT-OBJECT)))

(STOP-SCANNING
IF
((GOAL DO MENU TASK)
   (STEP VISUAL-SWEEP)
   (WM TARGET-TEXT IS ?T)
   (VISUAL ?TARGET-OBJECT LABEL ?T))
THEN
((DELDB (STEP VISUAL-SWEEP))
   (ADDDB (STEP MOVE-GAZE-AND-CURSOR-TO-TARGET))
   (ADDDB (WM TARGET-OBJECT IS ?TARGET-OBJECT))))

(MOVE-GAZE-AND-CURSOR-TO-TARGET
IF
((GOAL DO MENU TASK)
   (STEP MOVE-GAZE-AND-CURSOR-TO-TARGET)
   (WM TARGET-OBJECT IS ?TARGET-OBJECT)
   (WM CURSOR IS ?CURSOR-OBJECT)
   (MOTOR OCULAR PROCESSOR FREE)
   (MOTOR MANUAL MODALITY FREE))
THEN
((DELDB (STEP MOVE-GAZE-AND-CURSOR-TO-TARGET))
   (ADDDB (STEP MAKE RESPONSE))
   (SEND-TO-MOTOR OCULAR MOVE ?TARGET-OBJECT)
   (SEND-TO-MOTOR MANUAL PERFORM
      POINT RIGHT ?CURSOR-OBJECT ?TARGET-OBJECT)))
```

processor produces them very rapidly, as already described. Before the eye moves on, the digit in the fovea gets started in the perceptual recognition "pipeline" already mentioned, and eventually the recognized LABEL property of the object gets deposited in visual working memory. Meanwhile, the rule STOP-SCANNING functions as an independent "demon" waiting for an item to appear in visual working memory that matches the target. As soon as it does, this rule shuts down the scanning process by removing the item (STEP VISUAL-SWEEP) and then enables the rule MOVE-GAZE-AND-CURSOR-TO-TAR-GET. This rule then launches an eye movement and mouse cursor movement to the matching object. Thus, the search for the matching item is conducted partially in parallel, because the process of recognizing the labels for the desired object is overlapped in time with the eye movements required to move the fovea between the objects. Using the same parameters for the digit-recognition availability and time, this overlapping-search model predicts the values shown in Figure 3; the model predicts a slope of about 103 msec, which is an excellent match for the empirical value.

Implications for Eye and Hand Movements

This model makes claims about eye movements that might seem extreme and that certainly are subject to empirical test. Also, this is not the only possible EPIC model that could fit the data well, and there are additional effects in Nilsen's (1991) data to account for (see Hornof & Kieras, 1997). However, it is important to see how the claims of this model follow from the EPIC architecture. Because the menu items are uniformly spaced along a single line, the eye movements between the items are repeated movements, meaning that the feature structure of each movement is identical to that of the previous movement. Once started, the eye movements are very fast because the oculomotor-processor movement-preparation time is zero, leaving only the execution time. The cognitive processor simply commands each movement as soon as the oculomotor processor is ready. After a visual stimulus has been foveated, it can continue to be processed in the perceptual recognition pipeline while the eye moves on to the next item. The fully parallel capability of EPIC's cognitive processor makes it a simple matter to completely overlap the execution of the scanning process with the detection of the sought-for item. Thus, one possibility that EPIC presents is that the relatively fast menu selection time can be a simple result of scanning being done in parallel with matching.

Another possible model would be based on the idea that more of the menu items can be recognized outside the fovea (or the "effective" fovea is larger). For example, if about three digits can be discriminated simultaneously, then the scanning strategy could simply move the eye in jumps large enough to include each item in the "effective" fovea only once. This alternative model requires fewer, and less frequently executed, eye movements but still produces the overall fast menu selection times. In addition, there is another important effect in Nilsen's (1991) unordered menu data: Shorter menus are processed more rapidly across the board, which suggests a random search strategy like that argued for by Card (1984). By assuming that subjects perform a mixture of these different strategies, Hornof and Kieras (1997) were able to account for these effects in Nilsen's data. The goal of this line of work is a unified account of menu selection in terms of visual search and mouse pointing mechanisms, with the ultimate goal being an evaluation tool for visual layout: Two designs for the visual layout of the interface can be specified, and the corresponding EPIC models for the task could be run to determine which interface demands more visual work in the form of eye movements or perceptual delays.

An important detail in these results concerns the contribution of the mouse-movement time, which is well known to follow Fitts' law, a nonlinear function. However, both the predicted and observed times in these results are rather linear—what happened to the mouse-movement time? One possibility is that Nilsen's (1991) subjects did not make single mouse movements to the item but instead "scanned" the mouse cursor down the menu, halting at the desired item, as suggested by Sears and Shneiderman (1994). However, proposing that the eye could scan this rapidly is difficult enough to accept; that the hand could be moved this quickly seems quite implausible. A better and simpler explanation is the quantitative explanation provided by EPIC. It happens that over the range of target sizes and distances involved in Nilsen's menus, the Fitts'-law times for mouse movement are both relatively short compared to the total item-selection times and are also very close to being linear. Consequently, the mouse-movement time component of the total time does not result in noticeable nonlinearity. Thus, in contrast to verbal reasoning about qualitative properties of effects, detailed quantitative models such as EPIC can greatly clarify how different component mechanisms actually contribute to task-execution time.

Conclusions

This work illustrates how even an elementary aspect of using an interface can involve important and subtle aspects of the human performance architecture. Also, it shows how the quantitative parameter values built into EPIC impose powerful constraints on what strategies can serve as accurate models of cognitive processing. For example, it is possible to rule out the serial-search strategy because EPIC determines that the absolute minimum times for moving the eye and making the decision for each item leave no time for perceptual processing of the items. This minimum time, in turn, is determined by EPIC's feature representation of movements, in which the similarity of the repeated eye movements resulting from the

Figure 5. Production rule that enters each digit in the model for the telephone operator task. The output from the auditory recognition identifies the key to be struck by the manual motor processor.

```
(*Enter-number*Get-next-digit
IF
((GOAL Enter number)
   (STEP Get next digit)
   (WM Next speech is ?prev)
   (AUDITORY SPEECH PREVIOUS ?prev NEXT ?next TYPE DIGIT CONTENT ?digit)
   (VISUAL ??? SHAPE FIVE-KEY)
   (MOTOR MANUAL PROCESSOR FREE))
THEN
((SEND-TO-MOTOR MANUAL PERFORM Peck ?digit)
   (DELDB (WM Next speech is ?prev))
   (ADDDB (WM Next speech is ?next))))
```

specific visual layout of the task means that they will take little time to prepare. A final quantitative result provided by EPIC's mechanisms concerns the linearity of the selection-time function: The nonlinear component contributed by the mouse-movement time is not sufficiently large, given the specific layout of the menu, to produce a detectably nonlinear selection-time function.

3.2. Auditorily Driven Keyboard Data Entry

Another illustration that even elementary HCI tasks can involve considerable parallelism appears in modeling telephone operator tasks. A skilled telephone operator can listen to a string of digits spoken by a customer and enter them on a keypad while they are still being spoken. During such processing, the auditory perceptual processor, the cognitive processor, and the manual motor processor are operating simultaneously; the cognitive processor plays the role of mediator between the auditory and motor processors, feeding instructions to the motor processor as rapidly as the recognized digits become available from the perceptual processor and as soon as the motor processor is ready to accept instructions for another keystroke movement preparation. As part of Kieras et al.'s (1997) work, this performance was examined and modeled in detail.

Overlapping Auditory and Manual Processing

The auditory processor produces a series of auditory working memory items that contain the recognized digits recoded as identities of the keypad keys, chained together to preserve the order in which they were heard. Figure 5 shows a production rule, *Enter-number*Get-next-digit, from a model used in Kieras et al. (1997). The rule uses these recoded digits to make the corresponding keystrokes in a "pipeline" fashion similar in spirit to John's (1996) model of transcription typing. Each recognized spoken digit is represented in auditory working memory as a sequentially tagged item of the form (AUDITORY SPEECH PREVIOUS ?prev NEXT ?next TYPE DIGIT CONTENT ?digit), where the variable ?digit represents the recoding supplied by the auditory perceptual processor that designates the physical target of the corresponding key.

As each digit arrives in working memory, the rule fires when the manual motor processor has begun executing the previous keystroke, then sends the keystroke command corresponding to the digit to the manual motor processor, and also updates a "pointer" in WM to the next speech item in auditory working memory to be processed. The rule also requires that before the digit can be typed, the shape of the center key on the keypad, the FIVE-KEY, must be in visual working memory to ensure that the target key is in view. The manual motor processor uses the code for the digit key to prepare the features for a peck-style movement to strike the key and then initiates the prepared movement as soon as the hand is physically free to do so. The preparation process takes 0 to 100 msec, depending on the similarity of the present movement to the previous movement, while movement itself takes 50 msec to initiate, followed by a minimum of 100 msec to physically execute. Thus, there is enough time during each keystroke execution to prepare the features for the next keystroke, assuming that the cognitive processor has provided the next keystroke instruction soon enough. If so, then a sequence of keystrokes can be made quite rapidly, on the order of 150 to 200 msec apart. However, if the auditory input is not supplied rapidly enough, the keystrokes will be slower, but exactly how much slower depends on the subtle details of the event timing.

Buffering Effects

In the telephone operator task, the subtask of entering spoken digits is part of a larger task in which the operator must determine from the customer's speech what keys to strike to indicate the billing category of the call and then enter the billing numbers. The operator first indicates the billing category by striking the STA-SPL-CLG key, then strikes the KP-SPL key to signal that the billing number is about to be entered on the numeric keypad, and, then, enters the billing number digits. Performance in the whole telephone operator task was modeled by Kieras et al. (1997), but a special-ized EPIC model was used to explore some of the details of the auditory digit-entry subtask. In this model, the operator waits for the customer to speak the first digit before striking the STA-SPL-CLG key, followed by the KP-SPL key, and then strikes the digit keys according to the rule shown in Figure 5.

Individual stimulus and response timings were obtained from some of the videotaped performances used in Kieras et al. (1997). Figure 6 shows

Figure 6. Observed and predicted RTs for keystrokes made in response to speech input. Observed RTs are solid points and lines; predicted RTs are open points and dotted lines. The keystroke events are for four different task instances performed by a single subject. Each task instance involves a different sequence of keystrokes made in response to a billing request followed by digits spoken in a unique timing pattern. The keystrokes are shown in temporal order on the horizontal axis.

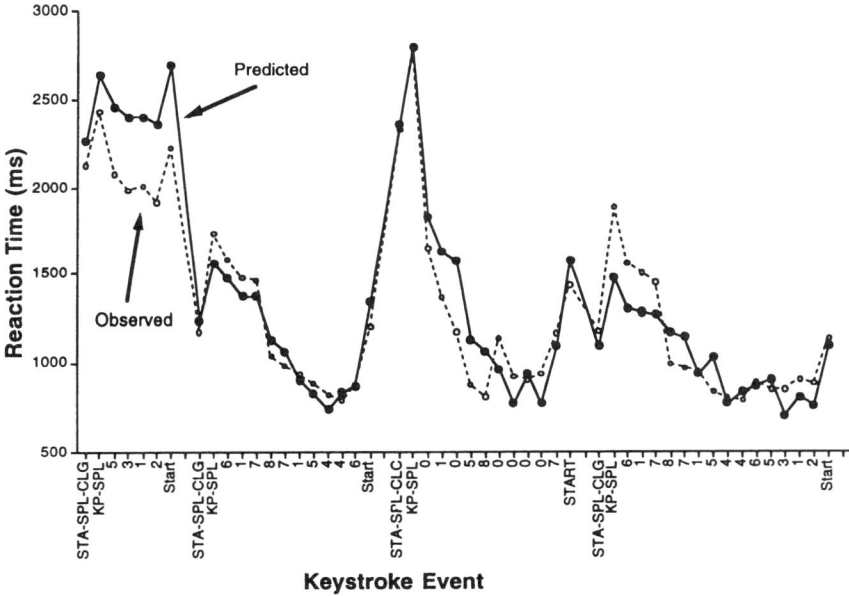

the observed and predicted RTs for each keystroke, measured from the auditory stimulus event defined as the stimulus for that keystroke. The observed RTs are from a set of four task instances performed by a single operator, where each task instance was a unique interaction between a customer and an operator, with a different sequence of digits, spoken in a different timing pattern, from the other instances. Thus, the RTs shown in Figure 6 are unaggregated, individual subject observations. Although there is a tendency for the RT to decrease during the sequence of keystrokes within a task instance, the RTs during this seemingly trivial task vary tremendously, both within and between task instances.

The values predicted from the EPIC model capture the general trend that the first several keystrokes are substantially delayed by the need for the previous keystrokes to be completed before the first digit keystroke. Gradually, the digit keystrokes catch up because the customer is speaking the digits at a lower average rate than they can be typed. The shortest RTs occur when the processing has caught up to the point that there is no idle waiting time. The auditory recognition parameter can be estimated as the difference between these observed shortest RTs and the predicted time for the cognitive and motor processing to produce these keystrokes. This parameter estimate, 400 msec, together with the earlier described strategy, suffices to produce the observed complex profile of RTs in response to the speech input. Given the single-observation quality of the data, the fit between predicted and observed RTs is quite good.

Conclusions

The model shows that many of the keystrokes in the task are delayed more than the architecture requires, suggesting that performance could be speeded up by changing the workstation design in two ways. First, if the first two keystrokes could be eliminated or placed elsewhere in the task, performance would be speeded up, and fewer digits would have to be buffered in working memory. Second, if the customer spoke the digits at a higher rate, or speech compression was used to produce the same effect, the task could in fact be done faster on the average. Given that, in this task domain, saving a second of average task time is credited with a considerable financial saving (Gray et al., 1993), the ability of EPIC to reveal these detailed aspects of performance is an important result.

3.3. Eye Movements and Executive Control in a Simple Dual Task

Computers are used not just in desktop or office settings but also in real-time contexts such as airplane cockpits, where many of the displays and controls are actually computer interfaces. As already mentioned, EPIC has been developed to deal with multiple-task situations, such as those in which a pilot has to track a target on one display and make decisions using the information in another display. The spatial distribution of visual information and the timing of eye movements play critical roles in determining performance in such situations. In the remainder of this article, we describe two studies, one involving a simple form of this paradigm and the other involving a complex form.

We have modeled some results obtained by Martin-Emerson and Wickens (1992) that are especially instructive about the role of eye movements and the use of visual information during dual tasks. Figure 7 shows the EPIC model display of the experimental task. The display represents the visual environment of EPIC with the objects in their correct sizes and positions; the small gray circle marks the location and size of EPIC's fovea, currently on the choice stimulus, and the larger gray circle marks the boundary of the parafovea, a region of intermediate discriminative ability.

The lower priority of the two tasks is a compensatory tracking task carried out in the upper box of the display. A quasirandom perturbing

Figure 7. EPIC model display for the Martin-Emerson and Wickens (1992) task. The tracking task is in the upper square; the eye is fixated on the choice-stimulus arrow appearing (not to scale) in the lower circle.

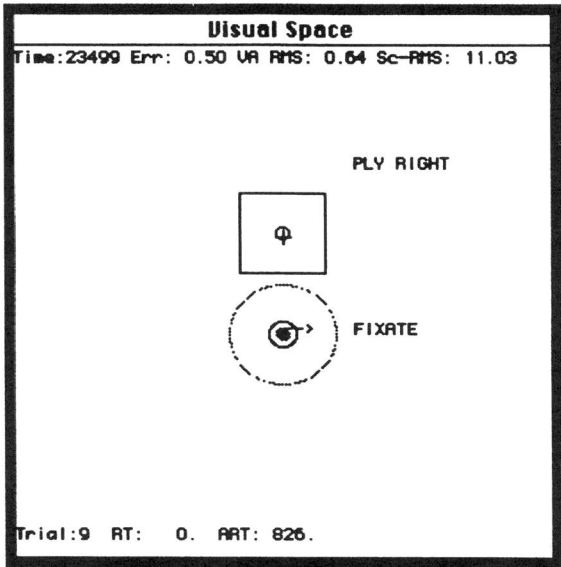

```
┌─────────────────────────────────────────┐
│             Uisual Space                 │
│ Time:23499 Err: 0.50 UR RMS: 0.64 Sc-RMS: 11.03 │
│                                          │
│                                          │
│                       PLY RIGHT          │
│                                          │
│                  ┌──────┐                │
│                  │  ϙ   │                │
│                  └──────┘                │
│                   ⌒ ⌒ ⌒                  │
│                 (  ◉➤ )  FIXATE          │
│                   ⌄ ⌄ ⌄                  │
│                                          │
│                                          │
│ Trial:9  RT:   0.  RRT: 826.             │
└─────────────────────────────────────────┘
```

force drives the cursor (the cross) away from the target (the small circle), and the subject must manipulate a joystick with the right hand to keep the cursor centered on the target. The higher priority of the two tasks is a choice-reaction task: Occasionally a stimulus appears in the choice-stimulus area (the solid circle below the tracking box), which is either a left- or right-pointing arrow (not shown to scale in the display). The subject must respond by pressing one of two buttons with the left hand as soon as possible, all the while attempting to maintain the cursor on the target.

The major independent variable is the distance (in visual angle) between the tracking target and the choice stimulus, and a second independent variable is the difficulty of the tracking task. The two dependent variables are the RT for the choice task and a measure of tracking performance (viz., the average root mean square error [*RMSE*] in the tracking task), collected for a 2-sec period following the onset of the choice stimulus. The observed effects are that the choice RT increases with the angular distance between the target and the choice stimulus but is unaffected by tracking difficulty. The *RMSE* increases somewhat with the angular distance for both levels of tracking difficulty.

Our models for this dual-task situation assume that successful tracking requires the eye to be kept on the tracking cursor, and likewise, the eye must be moved to the choice stimulus in order to discriminate it. However, if the choice stimulus is close enough to the tracking cursor, parafoveal vision will be adequate to discriminate the stimulus without moving the eye. Hence, the two tasks often, but not always, compete for use of the eye. Finally, because both tasks involve manual responses, they compete for access to the manual motor processor. We illustrate how EPIC can be applied to this task with two models.

A Simple Lockout Model of Executive Control

The lockout model uses a simple strategy that is consistent with traditional thinking about dual-task situations; namely, the lower priority task is locked out (suspended) while the higher priority task is executed. The strategy is shown in Figure 8, where our attempt to portray parallel interactive processes in flowchart form is explained as follows. The flowchart on the left-hand side of Figure 8 diagrams the tracking task. The tracking-task process waits for a cursor movement, and then it simply makes a motor movement if the cursor is too far off the target and waits for another cursor movement. In addition, another very simple iterative subprocess ensures that the eye stays on the cursor in case the autonomous oculomotor mechanism fails to keep up. On the right-hand side of Figure 8 is the flowchart for the choice task. When the stimulus appears, there may be a delay while an eye movement is made to it, followed by recognition of the stimulus and selection and production of a response. The executive process is the flowchart in the center; it monitors and controls the other processes via items in the production-system working memory; these relations are shown by the dashed arrows. The executive first starts the tracking task, allocates control of the eye to it, and then waits for the choice stimulus to appear. When the stimulus appears, the executive process suspends the tracking task, enables the choice-task rules, and then moves the eye to the stimulus if it is too far away to be discriminated. The executive waits for the choice response to be initiated, and then resumes the tracking task, and returns control of the eye to it. In this way, using what we term *lockout scheduling,* the executive process allows only one task to be done at a time, ensuring that the choice task has priority over the tracking task and that the eye and the manual motor processor are used for only one task at a time.

Unfortunately, this simple strategy does not fit all aspects of the data. Figure 9 shows the observed values for the choice RT and the tracking error together with the values predicted by the lockout model. By estimating the perceptual recognition time parameter from the data, we can fit the choice RT function fairly well. As in the data, there is no effect of tracking task difficulty because the choice task is given priority over tracking. The first few points are fairly flat, because here the choice stimulus can be recognized in the parafovea, and the eye does not need to be moved. The upward slope of the curves at larger separations reflects the time required

Figure 8. Flowchart of lockout-model strategy for the Martin-Emerson and Wickens (1992) task. The tracking task is suspended whenever the choice task is executing. The solid lines represent the flow of execution in each process; the dashed lines represent signal and control relations.

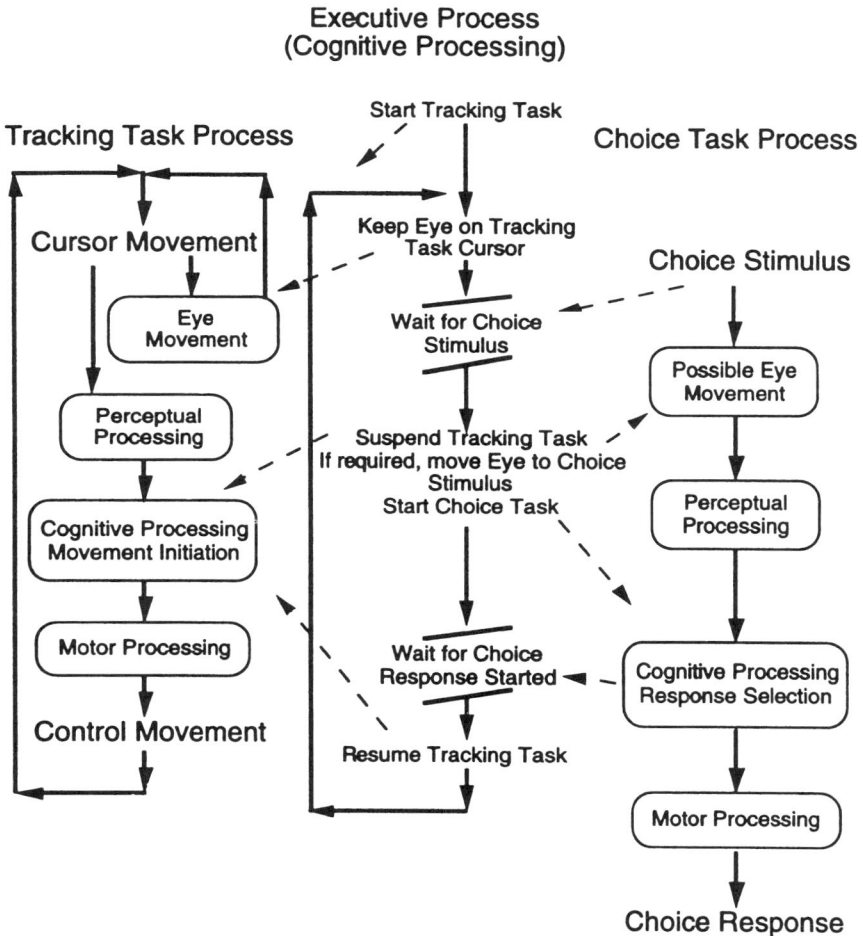

to move the eye. The fit of the simulated tracking data is extremely poor, however. The magnitude of the tracking error is seriously overpredicted, as are the effects of tracking difficulty and visual separation.

The lockout model cannot be made to fit the tracking data better by adjusting the relevant parameter values; it is already using the minimum plausible time estimates for all of the perceptual and motor parameters involved. Because the RMSE is measured for a brief (2-sec) period of time starting with the onset of the choice stimulus, if tracking is suspended for too long during this period, the effect will be substantial. The lockout

Figure 9. Observed and predicted effects of stimulus separation and tracking difficulty for the lockout model. Observed values are solid points and lines; predicted
values are open points and dotted lines. Square points are for the difficult-tracking
condition; circular points are for the easy-tracking condition. The fit of the choice RT
is satisfactory, but predicted tracking performance is far too poor.

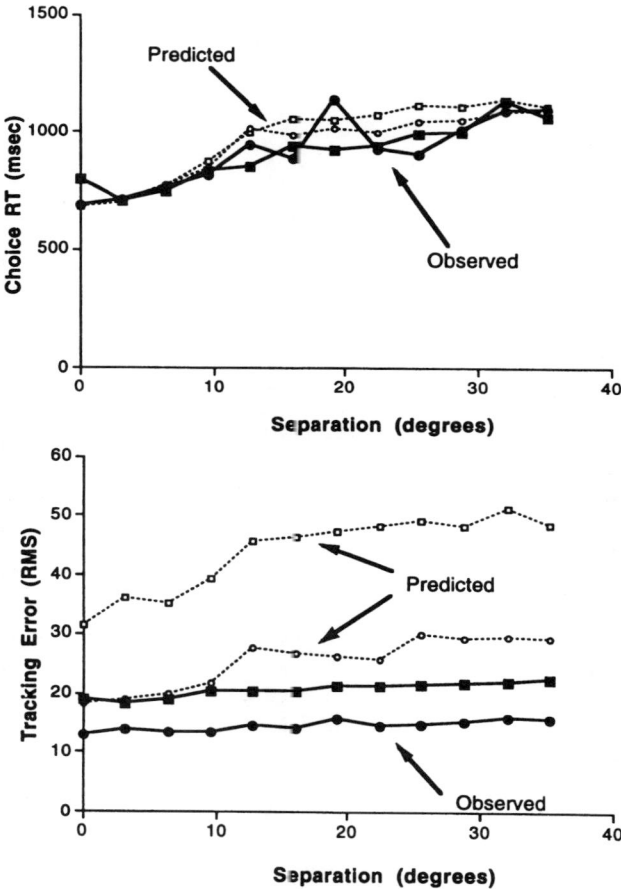

model suspends the tracking task for such a long time that considerable
tracking error accumulates; it is simply too inefficient.

An Interleaved Model of Executive Control

In order to provide a more efficient strategy, we constructed a second
model, the interleaved model, in which the executive process overlaps the
two tasks as much as possible; this strategy is shown in Figure 10. The
executive process starts out the same as in the lockout model, but when the

Figure 10. Flowchart of the interleaved model for the Martin-Emerson and Wickens (1992) task. The tracking task is executed while choice stimulus recognition and response selection go on and is interrupted for the minimum necessary time.

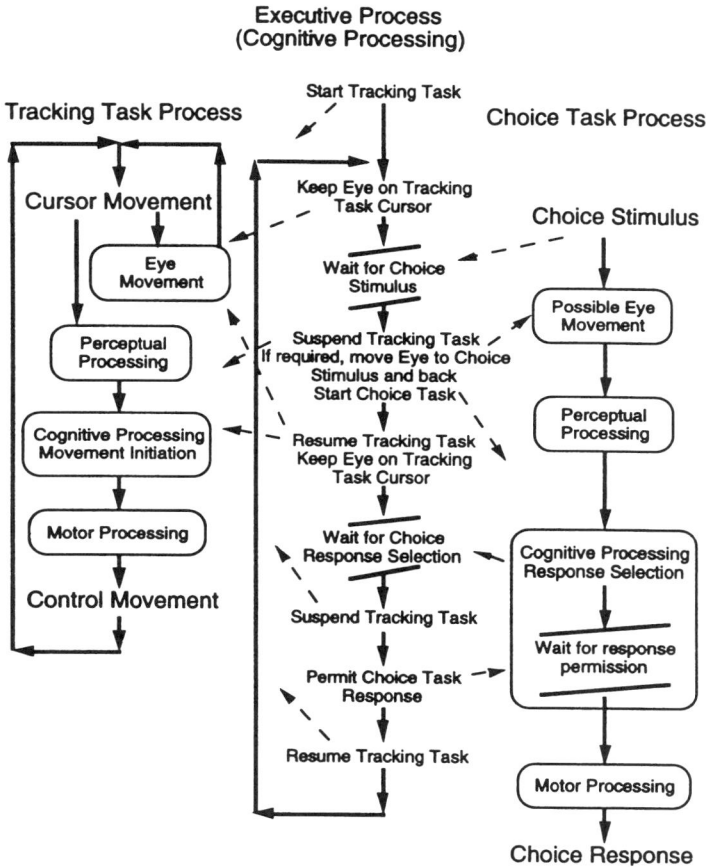

**Executive Process
(Cognitive Processing)**

choice stimulus appears, the executive moves the eye to it and then immediately begins to move the eye back to the tracking task, relying on the "pipeline" property of the visual system to acquire the stimulus and continue to process it even after the eye has returned to the tracking cursor. The tracking task is suspended only while the eye is away looking at the stimulus for the choice task. The executive then uses the same approach as in our PRP models for allocating control of the manual motor processor. When the choice task has chosen the response, it signals the executive, which again suspends the tracking task, gives the choice task permission to command the manual motor processor, and then resumes the tracking task right away. Thus, the same task priorities are honored, but the tracking task is interrupted as little as possible. The predictions from this model are shown in Figure 11. The choice RTs are again well fit, but now the tracking-task predictions are extremely close as well. Because the executive allocates the eye and the manual motor processor to the tracking task for the maximum amount of time, the tracking task rules can squeeze in a few movements while the choice task is underway, resulting in substantially better tracking error than in the lockout model.

Conclusions

As mentioned earlier, control of the eye has often been unappreciated, but it can clearly be critical in dual-task paradigms. A more subtle result is that subjects can and apparently do use highly refined strategies that can be surprisingly efficient for coordinating dual tasks. As an aside, in this model the executive handles the use of the eye directly, by moving it to the stimulus for the appropriate task. An alternative is to let each task move the eye itself, with the executive granting permission to move the eye to the appropriate task. This latter approach is the one followed in the complex dual-task model to be discussed.

An important general conclusion well illustrated by this particular modeling work concerns a common misunderstanding about computational models. They do not in fact have so many "degrees of freedom" that they can be made to fit any data at any time. Working within the fixed EPIC architecture sets powerful constraints. Given the basic lockout strategy, there are no parameter values or specific strategy details that would allow us to fit the data as a whole. The only way an EPIC model could fit the data is by assuming a fundamentally different strategy. Thus, a general conclusion (see also Meyer & Kieras, 1997b) is that the exercise of seeking quantitatively accurate accounts of data within a fixed architecture is extremely informative both about the accuracy of the architecture itself and about the structure and requirements of the task.

3.4. A Complex Dual Task With Automation

Our modeling work on the Martin-Emerson and Wickens (1992) task laid the foundations for our work on a more complex dual tracking/choice task. This task was developed by Ballas, Heitmeyer, and Perez (1992a, 1992b) to resemble a class of tasks performed in combat aircraft in which analyzing the tactical situation is partially automated by an on-board computer. To help with our explanation, we show the EPIC model display for this task in Figure 12. The right-hand box contains a pursuit-tracking task in which the cursor (cross) must be kept on the target (small box). In the experiment reported by Ballas et al., average tracking-error data were collected during various phases of the experiment. The left-hand box

Figure 11. Observed and predicted effects of stimulus separation and tracking difficulty for the interleaved model. Observed values are solid points and lines; predicted
values are open points and dotted lines. Square points are for the difficult-tracking
condition; circular points are for the easy-tracking condition. The fit is good for both
choice RT and tracking error.

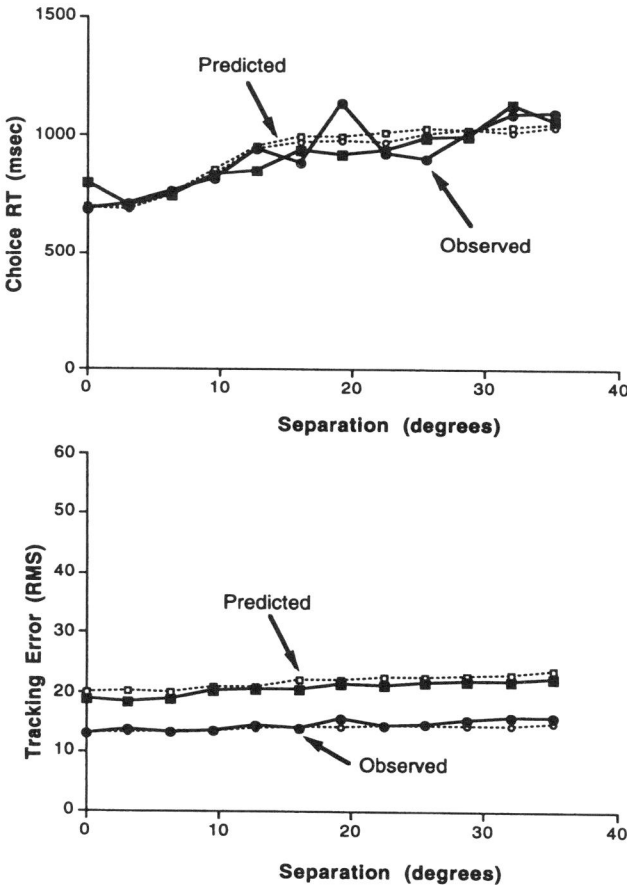

contains the choice task, a tactical decision task in which targets (or
"tracks") must be classified as hostile or neutral based on their behavior.
EPIC's eye is shown currently on one of the targets. These targets represent fighter aircraft, cargo airplanes, and missile sites that move down the
display as the subject's aircraft travels. Ballas et al. collected choice RT
data during performance of the tactical-decision task.

In the actual Ballas et al. (1992a) display, each type of target was coded
by an icon; for simplicity, in the EPIC display, they are represented

Figure 12. EPIC model display for the Ballas et al. (1992a, 1992b) task. The tracking-task target and cursor are on the right; four targets are moving down the tactical-decision task display on the left. The small open rectangle on the bottom left represents the response keypad, which has been displaced from its actual position for convenience; the small solid rectangle represents the current position of the finger used to "peck" keys. The eye has been positioned on the blue target, and one of the keystrokes is in progress.

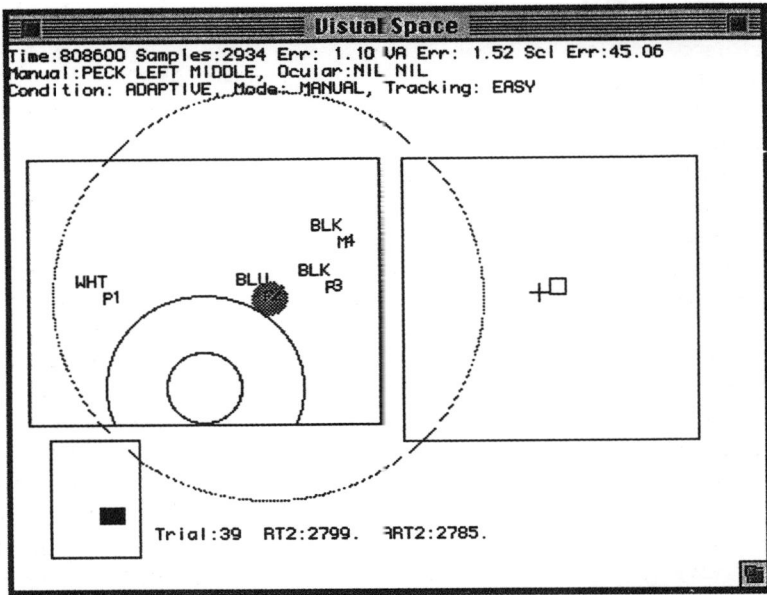

instead by a code letter. A track number identifies each object. Targets appear near the top of the display and then move down the display. After some time, the on-board computer attempts to designate the targets, indicating the outcome by changing the target color from black to red, blue, or amber, which the EPIC display shows with a three-character abbreviation. The subject must respond to the color change in a target in one of two ways. If the target becomes red (hostile) or blue (neutral) the subject must simply confirm the computer's classification; the response is striking a key for the hostile/neutral designation followed by a key for the track number. If the target becomes amber, the subject must classify the target based on a set of rules concerning the target's behavior and then type the hostility designation and track number. After the response, the target changes color to white and then disappears from the display some time later. The basic dependent variable is the two RTs for the targets, measured from when the target changes color to when the first and second of the two response keystrokes are made.

Ballas et al. (1992a, 1992b) investigated different interfaces for the tactical task. Our earlier description pertains to one of the four interfaces; the other three interfaces involved using a tabular display instead of the graphical radar-like display and a touchscreen instead of a keypad. The model presented here accounts for performance for the graphical-keypad interface already described. Additional work is underway on the other display and response formats.

A Performance Deficit Produced by Automation

Ballas et al. (1992a, 1992b) examined the effects of *adaptive automation.* These effects arise when, from time to time, the tracking task becomes more difficult, and the on-board computer takes over the tactical task, signaling as it does so. The computer then generates the correct responses to each target at the appropriate time, with the color changes showing on the display as in the manual version of the task. Later, the tracking becomes easy again, so the computer signals and then returns the tactical task to the subject; the experiment is arranged so that the subject must resume the tactical task when several black targets are on the display and one target has simultaneously changed color. Under such conditions, Ballas et al. observed an *automation deficit effect* in which, for a time after resuming the tactical task, subjects produced longer response times in the tactical task compared to their usual steady-state manual performance. This effect raises serious concerns about possible negative consequences of automation in combat situations; if the automation fails, the operator can lack *situation awareness,* and it might take a dangerously long time to catch up.

A Model for the Ballas Task

From single-task performance, we estimated the parameters for the basic perceptual encoding operations required in the tactical task, namely, recoding the blue and red colors to the appropriate key and recognizing the hostility of different kinds of targets, which takes considerably longer (more than a second). We have assumed that assessing the hostility of a target requires the target to be visually fixated, but that a target's color is available parafoveally, and that color-change events (which result in luminance changes) are visible in peripheral vision along with object onsets and offsets. Thus, when doing the tracking task, a color-change event in the tactical task can be detected, and this event can be used to tell that the tactical task requires action. However, like other transient events, this information will disappear from visual working memory quickly unless the task strategy involves recoding it into a more durable working memory form. After the eye has been moved to the tactical-task display, the colors of all of the individual targets will usually become available, because most often they will fall in the parafovea

Our model for this task and interface uses a lockout strategy at the top level of coordinating the tracking task and the tactical task but involves considerable parallelism within the tactical task. When the tactical task is the responsibility of the human (as opposed to that of the on-board computer), the executive process allows the tracking task to run until it is time to work on the tactical-decision task. In the meantime, the executive process simply notes that a target has appeared and continues tracking. The executive waits for a target to change color or to get too close to the centermost *ownship* circle; these events can be detected in peripheral vision, but the responsible target or color cannot be. The executive then suspends the tracking task and allocates the eye to the tactical task. The tactical task follows a priority scheme in choosing which target to view and process: Targets with a designation color (amber, red, or blue) are first priority, followed by targets whose color has changed, followed by targets whose color is unknown, and finally followed by an undesignated (black) target that is close to the ownship circle. If no targets qualify, the tactical task terminates, and the executive resumes the tracking task. If there is a qualifying target, the eye is moved to the chosen target; if there is more than one qualifying target at the same level of priority, one is chosen at random. The appropriate response is then made about the hostility of the target when the perceptual information (color coding or hostility behavior) becomes available; then the tactical-task process moves the eye to the target track number, and when the label is available, the corresponding response is chosen and made. After the second response is on its way to the manual motor processor, the process of choosing a new target begins in parallel with completion of the response. Thus, the tactical task has three major phases: choosing the stimulus to be processed and selecting and producing each of the responses for the chosen stimulus. Much of these three processes can be overlapped.

An Explanation for Automation Deficit

Our hypothesis about the source of the automation-deficit effect is that when resuming the tactical task, the tactical-task strategy must sort out a large number of targets, whereas during steady-state tactical-task operation, the targets are handled as they appear. That is, we assume that when the tactical task is automated, the subject does not bother to store any information about the state of the tactical display in working memory. Thus, when it is time to resume the tactical task, multiple targets are present on the tactical display, and there is no record in working memory of which have changed colors or when the color changes occurred. Thus, the information required to choose a target according to the priority

Figure 13. Automation deficit effect shown in the observed and predicted RTs for events following resumption of the tactical task. Observed RTs are solid points and lines; predicted RTs are open points and dotted lines. The lower curve is for the hostility-designation RT; the upper curve is for the track-number RT. Starting with Event 1, the events are closely spaced, and responses are delayed and then speed up as the tactical task catches up with the situation. Event 7 begins a similar matched sequence of closely spaced events, but the tactical task is performed as needed and so is able to keep up.

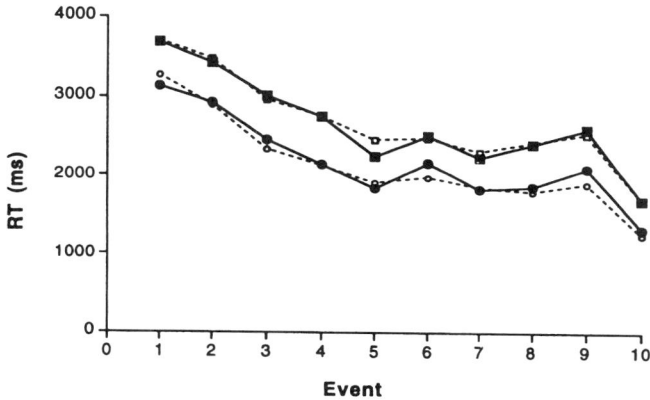

scheme is missing, so the tactical-task strategy simply picks the first target to inspect at random. After moving the eye to it and waiting for the color to become available, the strategy processes the target as usual if it is red, blue, or amber. However, if the target is white or black, it cannot be processed, and so another target is picked at random according to the priority scheme. When all candidate targets have been dealt with, tracking is resumed, and future target changes are processed as they appear.

The automation deficit results because when the tactical task is being performed normally, targets are usually processed in the order in which they change color, keeping the average RT to a minimum. In contrast, when the tactical task is resumed after automation, multiple targets must be inspected, and no information has been kept on the order in which they have appeared or changed color (otherwise, the automation is of little value!). Thus, the targets are inspected in random order, so targets that changed first will have to wait longer on average to be inspected than if they were processed as soon as they changed colors.

Figure 13 shows some automation deficit results obtained with the model in comparison to detailed data (supplied by James Ballas) that includes both RTs for correct responses only. The graph shows the predicted and observed RTs for each of the two responses for each target, measured from the time that they become designated (change color), in the order in which the targets become designated after the tactical task is resumed. Thus, Event 1 corresponds to the first color change of a target after task resumption, Event 2 to the second, and so forth. The overall quality of the fit is very good.

An important feature of these data is that the temporal spacing of events was systematically varied. Starting with Event 1, which is simultaneous with the signal to resume the tactical task, the events happen close together in time, and then thin out until by Event 6, usually only one target is present on the display; a matched pattern of event spacing starts with Event 7. Thus, the tactical processing is not evenly paced; there are two matched periods of high workload followed by easy stretches.

The Event 1 RT is long because the tactical task is started only after the auditory resumption signal has been recognized, and then it will frequently look at some other target first, further delaying the processing. Event 2 occurs very soon after Event 1, and so responding to it is also seriously delayed because it must wait for the delayed processing of Event 1 to be completed. The increasing event spacing allows the tactical task to gradually catch up, resulting in decreasing RTs. But starting with Event 7, the events are closely spaced again, and subsequent events suffer from the processing delays that again dissipate with the increasing event spacing. However, because the state of the display is being monitored at the time of Event 7, the tactical task is able to find its target much more quickly than is the case at Event 1. Thus, the effects of event spacing are more serious if the tactical-display monitoring is just being resumed than if it was ongoing. Ballas et al. (1992a, 1992b) defined the automation deficit effect as the difference between the Event 1 RT and the RT for a matched subsequent event, Event 7. The model accounts for this measure quite accurately.

Relation to Elementary Dual-Task Phenomena

Modeling this task has revealed a remarkable continuity with our earlier modeling work with EPIC on the PRP task mentioned earlier (Meyer & Kieras, 1997a, 1997b). The PRP task consists of two overlapping choice RT tasks; the major effect is that responding to the second task is delayed while the first response is being made. As the time between the two stimuli is increased, the RT to the second task declines toward its single-task value. We chose the PRP effect as the first phenomenon to address with EPIC because it was the simplest laboratory version of a dual-task paradigm and thus a good starting point. However, even Ballas et al.'s (1992a, 1992b) complex simulated-cockpit task produces PRP effects; that is, the declining RT pattern for the first six events in Figure 13 is a kind of PRP effect; the initial slow responses that gradually speed up as the stimulus spacing increases are due to exactly the same factors that govern the PRP effect in simpler laboratory paradigms. In other words, the automation deficit effect is a form of PRP effect. Our thorough understanding of PRP effects from the earlier modeling work has allowed us to account for performance of this realistically complex task in quantitative detail.

Conclusions

Our explanation for the automation deficit may have important implications for display and task design. For example, according to this hypothesis, resuming the tactical task could be done more efficiently if it is possible to easily select the highest priority object on the display at that time. That is, suppose the first-changed object currently on the display was coded by making it blink, which would be salient in peripheral vision. Then, the subject could simply look directly at the blinking object in order to ensure that the objects were processed in priority order. Alternatively, the automated version of the task could use a different, less salient way of representing its activity, so that the subject could still profitably monitor for the same perceptual events that are important in the manual version. Not only does the EPIC architecture supply a theoretical framework in which such issues can be explored and resolved in rigorous detail, but EPIC models can also be used to evaluate and predict the effects of the design changes implied by the explanations.

4. GENERAL CONCLUSIONS

EPIC is a computational architecture for constructing models of human cognition and performance that represent the contributions and interactions of perceptual and motor mechanisms as well as cognition in determining the time course of task execution. The examples presented in this article illustrate how EPIC can be applied to a variety of situations in which humans interact with computers both at the level of elementary interactions (e.g., menu operation and data entry) and at the level of high-speed concurrent execution of multiple display-intensive tasks. By accounting for empirical data with high accuracy in an architectural framework, models constructed with EPIC provide explanations for task phenomena with a clarity and precision far beyond the informal theorizing usually deployed in the HCI field. Further work with EPIC should lead to a comprehensive and predictive theoretical account of human performance in complex high-performance multimodal tasks.

At this point, EPIC is a research system that is not in a form suitable for routine use by system or interface designers. However, there is a technology transfer precedent: Earlier work with the CCT production rule models for HCI (Bovair et al., 1990) led to a practical interface design technique (John & Kieras, 1996; Kieras, 1988). Likewise, as the EPIC architecture stabilizes, and experience is gained in applying it to human–system analysis problems, we should be able to devise a simplified approach that will enable designers to apply EPIC to develop improved human–system interfaces.

Our experience with the EPIC architecture also suggests some meta-level conclusions about the role of cognitive modeling in the science and engineering fields of human performance and human–system interaction:

- Computational models based on human information-processing theory can usefully predict details of human performance in system design and evaluation situations.
- Developing and applying a cognitive model to task situations relevant to real design problems is a demand test of cognitive theory; if the theory successfully represents important properties of human abilities, it should in fact be useful in practical settings.
- Powerful constraints are imposed by quantitatively fitting fixed-architecture models to detailed performance data, which lead to the discovery of plausible task strategies.

In short, a comprehensive, detailed, and quantitative theory of human cognition and performance is the best basis for applied cognitive psychology. Rather than relying only on general psychological principles or on brute-force application of experimental methodology, system design can be best informed by using a theory that addresses phenomena at the same level of detail as design decisions require.

NOTES

Acknowledgments. Thanks are due to James Ballas of the Naval Research Laboratory for his generous assistance in making his task software available and in supplying new analyses of his data.

Support. This work was supported by Office of Naval Research Cognitive Sciences Program Grant N00014-92-J-1173 to David E. Kieras and David E. Meyer and by the Advanced Research Projects Agency (under Order B328 to David E. Kieras).

Authors' Present Addresses. David E. Kieras, Artificial Intelligence Laboratory, Electrical Engineering and Computer Science Department, University of Michigan, Advanced Technology Laboratory Building, 1101 Beal Avenue, Ann Arbor, MI 48109–2110. E-mail: kieras@eecs.umich.edu. David E. Meyer, Department of Psychology, University of Michigan, 525 East University, Ann Arbor, MI 48109–1109. E-mail: demeyer@umich.edu.

HCI Editorial Record. First manuscript received December 1995. Revision received December 1996. Accepted by Wayne D. Gray. Final manuscript received April 24, 1997. — *Editor*

REFERENCES

Anderson, J. R. (1976). *Language, memory, and thought*. Hillsdale, NJ: Lawrence Erlbaum Associates, Inc.

Anderson, J. R. (1993). *Rules of the mind*. Hillsdale, NJ: Lawrence Erlbaum Associates, Inc.

Ballas, J. A., Heitmeyer, C. L., & Perez, M. A. (1992a). *Direct manipulation and intermittent automation in advanced cockpits* (Technical Report NRL/FR/5534-92-9375). Washington, DC: Naval Research Laboratory.

Ballas, J. A., Heitmeyer, C. L., & Perez, M. A. (1992b). Evaluating two aspects of direct manipulation in advanced cockpits. *Proceedings of the CHI'92 Conference on Human Factors in Computing Systems,* 127–134. New York: ACM.

Bovair, S., Kieras, D. E., & Polson, P. G. (1990). The acquisition and performance of text editing skill: A cognitive complexity analysis. *Human–Computer Interaction, 5,* 1–48.

Card, S. K. (1984). Visual search of computer command menus. In H. Bouma & D. G. Bouwhuis (Eds.), *Attention and performance X: Control of language processes* (pp. 97–108). London: Lawrence Erlbaum Associates, Inc.

Card, S. K., Moran, T. P., & Newell, A. (1983). *The psychology of human–computer interaction*. Hillsdale, NJ: Lawrence Erlbaum Associates, Inc.

Elkind, J. I., Card, S. K., Hochberg, J., & Huey, B. M. (Eds.). (1989). *Human performance models for computer-aided engineering* (National Research Council, Committee on Human Factors). Washington, DC: National Academy Press.

Gopher, D. (1993). Attentional control: Acquisition and execution of attentional strategies. In D. E. Meyer & S. Kornblum (Eds.), *Attention and performance XIV: Synergies in experimental psychology, artifical intelligence, and cognitive neuroscience* (pp. 299–322). Cambridge, MA: MIT Press.

Gopher, D., & Donchin, E. (1986). Workload: An examination of the concept. In K. R. Boff, L. Kaufman, & J. P. Thomas (Eds.), *Handbook of perception and human performance, Vol. II: Cognitive processes and performance* (pp. 41.1–41.49). New York: Wiley.

Gray, W. D., John, B. E., & Atwood, M. E. (1993). Project Ernestine: A validation of GOMS for prediction and explaining real-world task performance. *Human–Computer Interaction, 8,* 237–309.

Hallett, P. E. (1986). Eye movements. In K. R. Boff, L. Kaufman, & J. P. Thomas (Eds.), *Handbook of perception and human performance* (Vol. 1, pp. 10.1–10.112). New York: Wiley.

Hornof, A. J., & Kieras, D. E. (1997). Cognitive modeling reveals menu search is both random and systematic. *Proceedings of the CHI'97 Conference on Human Factors in Computing Systems,* 107–114. New York: ACM.

John, B. E. (1996). TYPIST: A theory of performance in skilled typing. *Human–Computer Interaction, 11,* 321–355.

John, B. E., & Kieras, D. E. (1996). The GOMS family of user interface analysis techniques: Comparison and contrast. *ACM Transactions on Computer–Human Interaction, 3,* 320–351.

Kieras, D. & Meyer, D. (1995). Predicting human performance in dual-task tracking and decision making with computational models using the EPIC architecture. *Proceedings of the First International Symposium on Command and Control Research and Technology*, 314–325. Washington, DC: National Defense University.

Kieras, D. E. (1988). Towards a practical GOMS model methodology for user interface design. In M. Helander (Ed.), *Handbook of human–computer interaction* (pp. 135–158). Amsterdam: North-Holland Elsevier.

Kieras, D. E., Wood, S. D., & Meyer, D. E. (1997). Predictive engineering models based on the EPIC architecture for a multimodal high-performance human–computer interaction task. *ACM Transactions on Computer–Human Interaction, 4*, 230–275.

Kristofferson, A. B. (1967). Attention and psychophysical time. In A. F. Sanders (Ed.), *Attention and performance* (pp. 93–100). Amsterdam: North-Holland.

Laird, J., Rosenbloom, P., & Newell, A. (1986). *Universal subgoaling and chunking*. Boston: Kluwer.

Lee, E., & MacGregor, J. (1985). Minimizing user search time in menu retrieval systems. *Human Factors, 27*, 157–162.

Martin-Emerson, R., & Wickens, C. D. (1992). The vertical visual field and implications for the head-up display. *Proceedings of the 36th Annual Symposium of the Human Factors Society*, 1408–1412. Santa Monica, CA: Human Factors Society.

McMillan, G. R., Beevis, D., Salas, E., Strub, M. H., Sutton, R., & Van Breda, L. (1989). *Applications of human performance models to system design*. New York: Plenum.

Meyer, D. E., & Kieras, D. E. (1997a). A computational theory of executive cognitive processes and multiple-task performance: Part 1. Basic mechanisms. *Psychological Review, 104*, 3–65.

Meyer, D. E., & Kieras, D. E. (1997b). A computational theory of executive cognitive processes and multiple-task performance: Part 2. Accounts of psychological refractory-period phenomena. *Psychological Review, 104*, 749–791.

Newell, A. (1990). *Unified theories of cognition*. Cambridge, MA: Harvard University Press.

Nilsen, E. L. (1991). *Perceptual-motor control in human–computer interaction* (Technical Report 37). Ann Arbor: University of Michigan, Cognitive Science and Machine Intelligence Laboratory.

Norman, D. A., & Shallice, T. (1986). Attention to action: Willed and automatic control of behavior. In R. J. Davidson, G. E. Schwartz, & D. Shapiro (Eds.), *Consciousness and self-regulation* (Vol. 4, pp. 1–18). New York: Plenum.

Rosenbaum, D. A. (1980). Human movement initiation: Specification of arm, direction, and extent. *Journal of Experimental Psychology: General, 109*, 475–495.

Rosenbaum, D. A. (1991). *Human motor control*. New York: Academic.

Sears, A., & Shneiderman, B. (1994). Split menus: Effectively using selection frequency to organize menus. *ACM Transactions on Computer–Human Interaction, 1*, 27–51.

HUMAN–COMPUTER INTERACTION, 1997, Volume 12, pp. 439–462

ACT–R: A Theory of Higher Level Cognition and Its Relation to Visual Attention

John R. Anderson, Michael Matessa, and Christian Lebiere
Carnegie Mellon University

ABSTRACT

The ACT–R system is a general system for modeling a wide range of higher level cognitive processes. Recently, it has been embellished with a theory of how its higher level processes interact with a visual interface. This includes a theory of how visual attention can move across the screen, encoding information into a form that can be processed by ACT–R. This system is applied to modeling several classic phenomena in the literature that depend on the speed and selectivity with which visual attention can move across a visual display. ACT–R is capable of interacting with the same computer screens that subjects do and, as such, is well suited to provide a model for tasks involving human–computer interaction. In this article, we discuss a demonstration of ACT–R's application to menu selection and show that the ACT–R theory makes unique predictions, without estimating any parameters, about the time to search a menu. These predictions are confirmed.

John R. Anderson is a cognitive scientist with an interest in cognitive architectures and intelligent tutoring systems; he is a Professor of Psychology and Computer Science at Carnegie Mellon University. **Michael Matessa** is a graduate student studying cognitive psychology at Carnegie Mellon University; his interests include cognitive architectures and modeling the acquisition of information from the environment. **Christian Lebiere** is a computer scientist with an interest in intelligent architectures; he is a Research Programmer in the Department of Psychology and a graduate student in the School of Computer Science at Carnegie Mellon University.

CONTENTS

1. INTRODUCTION

ACT–R, as originally developed by Anderson (1993), was a model of higher level cognition. That model has been applied to modeling domains like the Tower of Hanoi, mathematical problem solving in the classroom, navigation in a computer maze, computer programming, human memory, and other tasks. By the standards of these fields, it has provided good models of human cognition. However, by the standards of human–computer interaction (HCI), it has had a serious failing. It has ignored many of the details by which the subject interacted with the external environment. All of the applications, either in the laboratory (e.g., Anderson, Reder, & Lebiere, 1996) or in the classroom (e.g., Anderson, Corbett, Koedinger, & Pelletier, 1995), have involved people reading a computer screen and using a mouse and a keyboard, but there was no theory of how this "input" and "output" took place. In Kieras and Meyer's (1994) terms, we had a theory of "disembodied cognition." We have recently addressed these lacunae and have developed a theory of how ACT–R interacts with computer applications. Most of these embellishments are concerned with visual attention, although they also address issues of visual perception and motor action. We describe this added theory of visual attention and how it relates to the ACT–R theory of higher level cognition. We describe its application to some classic paradigms in visual attention to establish its credibility. Then we describe its extension to a menu-selection task and its ability to make some novel predictions about that task. First, though, we set forth the basic ACT–R theory of cognition. We are brief in our description of the basic ACT–R theory because descriptions exist elsewhere (Anderson, 1993). Here we just describe enough detail to establish the context in which our theory of visual attention has been developed.

ACT–R assumes that there are two types of knowledge—declarative and procedural. *Declarative knowledge* corresponds to things that we are aware we know and can usually describe to others. Examples of declarative knowledge include "George Washington was the first president of the United States" and "Three plus four is seven." *Procedural knowledge* is knowledge that we display in our behavior but of which we are not conscious. Procedural knowledge basically specifies how to bring declarative knowledge to bear in solving problems.

Declarative knowledge in ACT-R is represented in terms of chunks (Miller, 1956; Servan-Schreiber, 1991), which are schema-like structures consisting of an "isa" pointer specifying their category and some number of additional pointers encoding their contents. Below is a chunk encoding the addition fact that $3 + 4 = 7$:

```
fact3 + 4
    isa        addition-fact
    addend1    three
    addend2    four
    sum        seven
```

Production rules specify how to retrieve such declarative knowledge to solve problems. For instance, consider a child working on the 10s column in the following multicolumn addition problem:

$$
\begin{array}{r}
234 \\
+\ 746 \\
\hline
0
\end{array}
$$

At this point in time, the following production rule might apply:

```
IF the goal is to add n1 and n2 in a column,
    and n1 + n2 = n3
THEN set as a subgoal to write n3 in the column
```

Applied to the preceding problem, this production rule would retrieve the addition fact $3 + 4 = 7$ and set the subgoal to write out 7 in the 10s column. At this point, other productions would apply, which would deal with operations like carrying into or out of the column or writing out the answer.[1] All productions in ACT-R have this basic character of responding to some goal, retrieving information from declarative memory, and possibly taking some action or setting a subgoal. In ACT-R, cognition proceeds step by step by the firing of such production rules.

Other aspects of ACT-R involve a theory of subsymbolic, neural-like computations that determine the availability of declarative chunks and

---

1. For a complete model of multicolumn addition, see Anderson (1993, chaps. 1 & 2), where the formal syntax of such production rules is specified.

choice among production rules. These aspects of the theory are important in determining timing when retrieval from memory becomes important and in predicting which paths subjects will explore in complex problems. There is also a learning theory that specifies the acquisition of symbolic chunks and productions as well as subsymbolic continuous-valued quantities associated with the chunks and productions. This learning theory is critical to modeling skill acquisition. We do not elaborate on the subsymbolic or learning aspects of ACT–R because they are not critical to the tasks described here. What is critical is ACT–R's new theory of visual attention.

2. A THEORY OF THE VISUAL INTERFACE

Theories of higher level cognition typically ignore lower level processes such as visual attention and perception. They simply assume that lower level processes deliver into working memory some relatively high-level description of the stimulus situation upon which the higher level processes operate. This certainly is an accurate characterization of our past work on the ACT–R theory (e.g., Anderson, 1993). The typical task to which ACT–R has been applied is one in which the subject must process some visual array (the array may contain a sentence to be recognized, a puzzle to be solved, or a computer program to be written). We have always assumed that some processed representation of this visual array is placed into working memory in some highly encoded form, and we modeled processing given that representation.

The strategy of focusing on higher level processes might seem eminently reasonable for a theory of higher level cognition. However, the strategy creates two stresses for the plausibility of the resulting models. One stress is that, by assuming a processed representation of the input, the theorists are granting themselves unanalyzed degrees of freedom in terms of choice of representation. It is not always clear whether the success of the model depends on the theory of the higher level processes or on the choice of the processed representation. Another stress is that theorists may be ignoring significant problems in access to that information that may be contributing to dependent variables such as accuracy and latency. For instance, the visual input often contains more information than can be held in a single attentional fixation, and shifts of attention (with or without accompanying eye movements) may become a significant but ignored part of the processing. For these reasons, we have been encouraged to join several other efforts (e.g., Kieras & Meyer, 1994; Wiesmeyer, 1992) to embed a theory of visual processing within a higher level theory of cognition. The choice to focus on vision is largely strategic, reflecting the fact that most of the tasks that ACT–R has modeled have involved input from the visual modality. To be more exact, most tasks have involved processing input from a computer screen, and so we developed a theory of the processing of a computer screen. As a fortunate consequence, we have situated ACT–R to be appropriate for HCI applications.

We wanted to remove anything implicit about how our theory related to the behavior we saw from our subjects. To do this, we have our simulation interact with the same software that presents the experiment to the subject. Basically, our ACT–R simulation can operate the computer application just as a subject can: As we describe, the simulation has access to the same computer screens to which the subject has access, must scan these screens as a subject must, and must enter keystrokes and mouse motions as a subject must. The software does not distinguish whether the keystrokes and mouse motions come from ACT–R or a human. The data from the simulation are collected by the same software that analyzes the human's data and are subject to the same analyses. The one difference between our simulation and a human is that ACT–R's whole world is the computer screen, the mouse, and the keyboard, whereas this is only a small part of the human's world.

It is important to define our approach from the outset: We require a theory of visual attention and perception that is psychologically plausible, but it is not our intention to propose a new theory of visual attention or perception. Therefore, we have embedded within ACT–R a theory that might be seen as a synthesis of Posner's (1980) spotlight metaphor, Treisman's feature-synthesis model (Treisman & Sato, 1990), and Wolfe's (1994) attentional model. These seem to define the current consensus model of visual attention and perception. What this model does is provide us with a set of constraints that we can then embed within the ACT–R theory of higher level cognition.

Figure 1 provides a basic overview of the system. There are three basic entities to be related. There is the ACT–R system, which we have been describing; there is the environment with which the system is interacting (in our case, the computer application); and there is an iconic memory, which is a feature representation of the information on the screen.[2] As can be seen, there is a limited number of actions that ACT–R can take—it can issue keystrokes and mouse presses to the computer, and it can move its attention around the iconic memory. Wherever it moves its attention, it can synthesize the features located there into declarative chunks that can then be processed by the ACT–R system. The computer program with which it is interacting can issue updates to the screen (and thence to the iconic memory) either spontaneously or in response to actions of ACT–R.

---

2. Thus, what ACT–R "sees" when a word is presented is neither a pixel representation nor a word but a feature description of the word. This corresponds to what the visual cortex receives from lower level routines. ACT–R must recognize the patterns (i.e., words) formed by these feature descriptions.

Figure 1. Relation among ACT–R, the environment, and iconic memory.

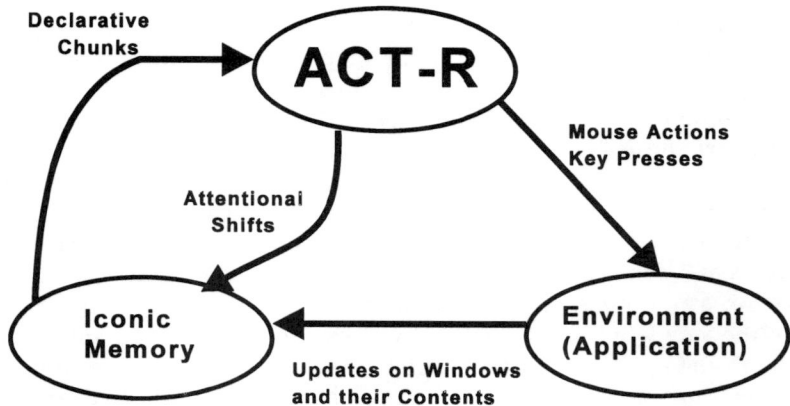

We flesh out this basic description in the subsequent sections of this article. In the next section, we focus on the theory of visual attention, which serves to guide attention in its movement about the screen. These are the kinds of cognitive operations that must be addressed in many HCI applications. After describing the theory of visual attention, we address some of the "classic" studies of visual attention. In the final section, we show how these ideas can model, with no additional assumptions, a canonical HCI task—selecting an item from a menu.

3. VISUAL ATTENTION

In processing information from a computer screen, we do not have constant access to everything and often have to search for information. On the other hand, we rarely have to do an exhaustive search of the entire screen to find what we are seeking. ACT–R's theory of visual attention is concerned with how ACT–R finds and extracts information from the iconic memory in Figure 1. The information in the visual icon consists of features, but ACT–R cannot process visual features directly. It can only process chunks representing the objects that these features compose. We have implemented the spotlight metaphor of visual attention, in which a variable-size spotlight of attention can be moved across the visual field. When the spotlight fixates on an object, its features can be recognized. Once recognized, the objects are then available as chunks in ACT–R's working memory and can receive higher level processing. The following is a potential chunk encoding of the letter *H:*

```
object
    isa H
    left-vertical      bar1
    right-vertical     bar2
    horizontal         bar3
```

We assume that, upon the appearance of an object in the visual field, the features comprising the object (e.g., the bars) are available but that the object itself is not immediately recognized. The system can respond to the appearance of a feature anywhere in the visual field. Only when it has moved its attention to that location can it recognize the conjunction of features that correspond to the object. Thus, for instance, it can respond immediately to a vertical bar but can recognize an *H* only after moving attention to that object. Thus, in order for the ACT–R theory of higher level processing to "know" what is in its environment, it must move its attentional focus over the visual field. In ACT–R, the calls for shift of attention are controlled by explicit firings of production rules. Consequently, it will take time to encode visual information, and we are forced to honor the limited capacity of visual attention.

What information can ACT–R use to guide where it looks on a screen? There are three basic types of information ACT–R can use to guide where attention goes: ACT–R can (a) look in particular locations and directions, (b) look for particular features, and (c) request to scan for objects that have not yet been attended. ACT–R can conjoin these in scanning requests, asking for things like, "Find the next unattended pink vertical bar to the left of the current location." This kind of search deserves several comments. First, note that ACT–R can search for a conjunction of visual features (pink and vertical). At one time, it had been argued that attention could be drawn only by single features (e.g., Treisman & Gelade, 1980). However, a more current view is that attention can be guided by conjunctions of features but that such conjunction searches are more noisy (Wolfe, 1994).[3] Second, ACT–R can specifically restrict itself to unattended objects. There is evidence that people have difficulty returning attention to attended objects even if they want to (Klein, 1988; Tipper, Driver, & Weaver, 1991). Although ACT–R can restrict itself to unattended objects, it has no more difficulty attending to previously attended objects than to previously unattended objects. Thus, this "inhibition of return" is not modeled in the ACT–R visual component. At some point in time, we might extend ACT–R's attentional module to incorporate these details, but right now it should be viewed as a system that is consistent at a general level with what we know about visual attention but that does not model the

---

3. However, ACT–R does not yet model this noise.

Figure 2. ACT–R can see either the *H* or the *X*s comprising the *H,* depending on how ACT–R sets its feature scale.

```
     X        X
     X        X
     X        X
     XXXXXX
     X        X
     X        X
     X        X
```

microstructure of these attentional processes. As stated earlier, our goal is to focus on how visual attention is used by the cognitive system.

A final general comment is that ACT–R can select the scale of the features for which it searches and the size of the object it is recognizing. Thus, it can recognize either letters or words as objects. Also, depending on how ACT–R sets its feature scale, we would want it to recognize (in Figure 2) either the *H* or the *X*s comprising the *H*. In fact, subjects can adjust the scale or the spatial frequency at which they are attending to a visual display (Navon, 1977).

The best way to understand how this theory works is to see it applied to various tasks involving visual attention. A constant problem in processing a visual array, such as a computer screen, is finding objects on that screen—whether one is looking for an icon, searching a menu, or scanning a text for a key word. Such visual-search tasks have been a bread-and-butter domain of experimental research on visual attention. There is a rich literature surrounding such tasks, and we want to establish that the ACT–R theory of visual attention is consistent with this literature. We look at three classic paradigms–the Sperling paradigm, subitizing, and speeded search. At one level, the ACT–R theory just implements the existing theoretical understandings of these paradigms. However, the ACT–R implementation does so in precise mechanistic terms and so banishes the homunculus that tends to haunt theories of visual attention. Also, because it places all of these tasks into a single framework, ACT–R allows us to establish that there is an apparent universal of visual attention, which is that it takes about 185 msec to move visual attention. With this parameter established, we are then able to make novel a priori predictions, free of additional parameters, about menu search.

3.1. Sperling Task

Sperling (1960) reported a now-classic study of visual attention. Figure 3 illustrates the material Sperling used in one of his experiments. In the whole-report condition, he presented subjects with brief presentations (50

Figure 3. An example of the kind of display used in a visual-report experiment. This display is presented briefly to subjects, who are then asked to report the letters it contains.

X	M	R	J
C	N	K	P
V	F	L	B

Figure 4. Data from Sperling (1960). Number of items reported from a row of four items as a function of the delay of the cue identifying the row.

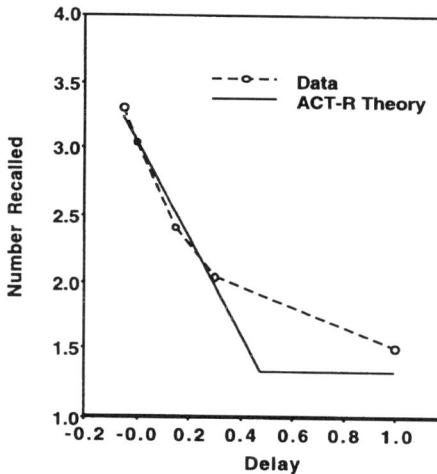

msec) of visual arrays of letters (three rows, four columns) and found that, on average, subjects could report back 4.4 letters. In the partial-report condition, Sperling gave subjects an auditory cue to identify which row they would have to report. He found that they were able to report 3.3 letters in that row. When he delayed the presentation of the auditory cue to 1 sec after the visual presentation, he found that subjects' recall fell to about 1.5 letters. Figure 4 shows Sperling's results as a function of the delay in the tone. Because subjects' recall at 1 sec of delay fell to about one third of the whole-report level, the obvious interpretation was that subjects were able to report as many items from the cued row as they happened to encode without the cue. This research has been interpreted as indicating that subjects have access to all of the letters in a visual buffer but that they have difficulty reporting them before the letters decay.

This experiment and other subsequent research have two dimensions of significance. The first dimension is information about the limitations of visual sensory memory. The general importance of this limitation has been questioned (e.g., Haber, 1983), and it certainly does not seem significant to

HCI issues for which we do not receive 50-msec screen presentations. The second dimension of significance is how fast visual attention can move over an array, which is quite relevant to many domains, including the processing of computer screens. For this reason, we believe it is important to show that the ACT–R theory of visual attention can model this result.

We developed a simulation of this task in which the letters in the visual array were encoded by the visual interface as sets of features grouped into unidentified objects. When a report row is not identified, the following production would apply:

Encode-Screen
IF one is encoding digits without a tone
and there is an unattended object on the screen
THEN move attention to that object

After a row has been identified, different productions would fire depending on the tone. For instance, the following production is responsible for reporting the top row:

Encode-Top-Row
IF one is encoding digits and there is a high tone
and there is an unattended object in the top row
THEN move attention to that object

These productions call for attention to be moved to unattended objects. When the production moves attention to the location of that object, the letter would be recognized and a chunk created to encode it. If no tone is presented, **Encode-Screen** will encode any letter in the array, whereas, if a tone is present, productions like **Encode-Top-Row** will encode letters in the cued row. After the visual array disappears, the following production is responsible for report:

Do-Report
IF the goal is to report the digits
and there is a chunk encoding an item
THEN report the item

This production will report only those letters that had been encoded because only these have chunk representations in working memory.

The number of letters encoded in the whole-report procedure is essentially equal to the number of **Encode-Screen** productions that can fire before the iconic memory of the letters disappears. Physically, the stimulus is presented for only 50 msec, but the critical issue is the duration of the stimulus in the system—a parameter we estimate to be 4.4 times the firing time per production (as 4.4 items are recalled on average). In fitting the data in Figure 4, we estimated the duration of the image to be 810 msec and the time per production to be 185 msec, which is a reasonable estimate for the time for attention to move. Note that 810 / 185 = 4.4.[4]

To understand the fit to the data in Figure 4, we needed to think through how the advantage of the partial report worked. In our analysis, subjects had a one-in-three chance of guessing the right row, in which case they would be able to report the four letters. They had a two-in-three chance of guessing wrong, in which case they would only start encoding the row after switching to that row. We assumed that there was some delay in time for the tone to be perceived and for attention to switch to the correct row (note in Figure 4 that subjects never reported all four items and were doing better given a .05-sec headstart on the tone than a simultaneous presentation). We estimated this switch-over delay to be 335 msec. This can be seen as 150 msec to register the signal (the time for auditory signal to get from the ear to being registered in the goal chunk) and 185 msec for an attention-changing production to fire (same time as value of all other attention-switching productions). Thus, the effective time spent encoding an array if the tone is presented t msec after the array will be $810 - t - 335$ msec. Thus, our predicted number of digits reported is:

$$\tfrac{1}{3} \times 4 + \tfrac{2}{3} \times \left[\frac{(810 - t - 335)}{185} \right] = 3.04 - .0036t \quad \text{if} \quad 810 - t - 335 > 0 \text{ (or } t < 475)$$

or

$$\tfrac{1}{3} \times 4 = 1.33 \qquad\qquad\qquad\qquad \text{if} \quad 810 - t - 335 < 0 \text{ (or } t > 475)$$

Figure 4 presents the predictions of this model. As can be seen, the model does a nice job of simulating the data. The ACT–R model of this task is very simple and consists of the production rules given plus a rule to switch from attending to reporting. In part, it is implementing the standard understanding of the data, but it makes clear both the cor'rol structure of the task (which is vague in the standard understanding) nd the need to postulate the switching time (335 msec) to consistently account for the data. For purposes of comparison with later modeling efforts, the critical number is 185 msec for switching attention. This number comes directly from the slope in Figure 4. Every 185 msec, the memory report is dropping by two thirds of an item.

---

4. An Excel file giving the fit can be found by following the path given in the Background note at the end of this article.

Figure 5. Data from Jensen, Reese, and Reese (1950). Amount of time to name the number of objects in a presentation as a function of the number of objects.

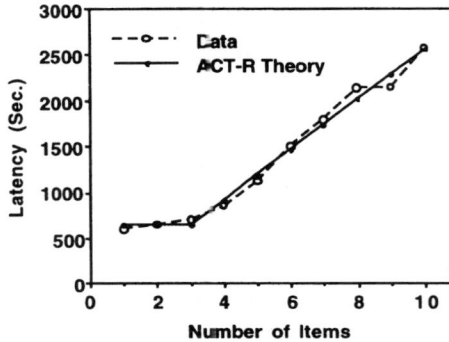

3.2. Subitizing Task

In the Sperling task, time is controlled by the duration of the iconic memory, and the goal is to see how many things can be attended to in that time. Another way to measure switching time for attention is to see how long it takes to attend to several objects on a screen. One way to get people to attend to all of the objects on a screen is to ask them to say how many objects there are. This is precisely what is done in a subitizing task (see the recent discussion by Simon, Cabrera, & Kliegl, 1994), in which several objects are presented to a subject, and the subject must identify as quickly as possible how many objects there are on the screen. Figure 5 illustrates the classic result obtained (Jensen, Reese, & Reese, 1950) in this task in which there is an increase in latency with number of digits to be identified. There is an apparent discontinuity in the increase, with the slope being much shallower until three or four items and then getting much steeper. There is about a 50-msec slope until three or four items and approximately a 275-msec slope afterward. Figure 5 also shows the results from the ACT–R simulation that we describe

The basic organization of the model is to assume that there are special productions that recognize one object, two objects (e.g., lines), three objects (e.g., triangles), and familiar configurations of larger numbers of objects (e.g., the five on a die face) and that there is a production that can count single objects. This is the basic model of the subitizing task that has been proposed by researchers such as Mandler and Shebo (1982). Again, what ACT–R adds to this standard model is an explicit theory of the control structure. The following is some of the productions used in modeling the task:

Start
>IF the goal is to count the objects starting from a count of 0
>
>THEN move attention to some object on the screen

See-One
>IF the goal is to count the objects starting from a count of 0
>   and a single object has been seen
>
>THEN initialize the count to 1

See-Two
>IF the goal is to count the objects starting from a count of 0
>   and a line of two objects has been seen
>
>THEN initialize the count to 2

See-Three
>IF the goal is to count the objects starting from a count of 0
>   and a triangle of three objects has been seen
>
>THEN initialize the count to 3

Attend-Another
>IF the goal is to count the objects and the count is not 0
>   and there is another unattended object
>
>THEN move attention to that object

Add-One
>IF the goal is to count the objects and the count is not 0
>   and another object has been attended
>   and X is one more than count
>
>THEN reset the count to X

Stop
>IF the goal is to count the objects
>   and there are no more unattended objects
>
>THEN respond with the count

Faced with an array of objects, **Start** will move attention to some part of the screen, and the largest pattern will be recognized. In this model, we have introduced the capacity to see patterns of one, two, and three objects. Depending on which pattern is attended to, one of the productions (**See-One**, **See-Two**, or **See-Three**) will apply to initialize the count. After that point, **Attend-Another** will move attention to other unattended objects, and **Add-One** will add one to the count. When there are no more unattended objects, **Stop** will report the count.

There are several noteworthy aspects of this model. First, it makes clear that successful performance of subitizing depends on ACT–R's ability to tag items in the visual array as attended so that double counts are avoided. Second, beyond three, subitizing depends on retrieval of counting facts. One could have an alternate model that aggregated additional items in units larger than one. Thus, six objects might be achieved by twice attending three objects and adding $3 + 3 = 6$. However, retrieval of such addition facts would be much slower than retrieval of counting facts. The model predicts a flat function from one to three and an equal rise from three to four as from four to five—neither of which is quite true. This may reflect the fact that there is some probability of counting in the "sub-three" range and some probability of pattern matching for four elements. Although the model could be complicated to incorporate these ideas, it did not seem worth it. The correlation between prediction and data was already .995.[5]

The most important issue with respect to coordinating this account with our model of the Sperling task is accounting for the 275 msec slope that holds beyond four digits. In fitting this data, we assumed a 185-msec time to switch attention, as in the Sperling model. However, ACT–R does predict the 275-msec slope because it takes approximately an additional 90 msec to retrieve the counting fact in production **Add-One**—x is one more than the count. Although this 90-msec period is estimated for this experiment, it is consistent with estimates in our model of cognitive arithmetic (Lebiere, 1997).

3.3. Visual-Search Task

Another way to investigate the time to shift attention is to display an array and ask subjects to search among objects for a specific object. If one can manipulate the number of objects through which a subject must search, one can manipulate search time. The slope of the function relating number of objects attended to to search time gives an estimate of time to move attention. This straightforward logic is complicated by the fact that subjects can select which objects to attend to on the basis of the features of the objects. Thus, for instance, in looking for a red object, subjects will not be affected by the number of green objects in the array.

An example reflecting such a paradigm and its complexities is Shiffrin and Schneider's (1977) study of visual search. In their Experiment 2, subjects had to detect a target item when it was presented in a visual display of one to four items (frame size). The target letter was in a memory set of one to four (memory-set size). For instance, subjects might hold a memory set of B and K and be asked if either element occurred in a visual

---

5. An Excel file giving the fit can be found by following the path given in the Background note at the end of this article.

array that contained G, K, M, and F (in which case, they would respond yes). Subjects were either in what was called the varied-mapping condition or what was called the consistent-mapping condition. In the varied-mapping condition, both distractors and the memory-set items were letters (drawn from the same pool on each trial); in the consistent-mapping condition, the memory set was composed of numbers, and the distractors were letters (therefore, they were always drawn from different pools). In general, judgment times increase with memory-set size and frame size, but the effects are much stronger for the varied-mapping condition. Figure 6 (top panel) shows Shiffrin and Schneider's results.

We developed what is a fairly straightforward ACT–R model of this task. It involved the following stages:

1. *Preparation*. Upon receipt of the memory set, an effort was made to find a feature common to all members of the memory set. If there was more than one such feature, then the feature was selected that was least frequent among the distractors. If no one feature characterized all of the items in the memory set, two or more features could be selected. This defined the target feature set. The feature set we used was the features proposed by McClelland and Rumelhart (1981) plus one global feature to encode whether the character was left-facing, right-facing, or symmetric.

2. *Search*. ACT–R directed attention to a location on the basis of the target set of features. Upon presentation of the display, the system examined all positions that had a target feature. If no position had a target feature, it randomly selected one position to view. It would look at more only if more than one position had a target feature. This search was self-terminating in the case of positive trials, but all positions with target features had to be examined in the case of negative trials. The first production that applies to start the scanning is:

Encode-First-Object   **(SHIFT)**
   If the goal is to search for an object with feature F
      and an unattended object with feature F occurs in location L
THEN move attention to that object

3. *Judgment*. For each position examined, it decided whether that item was in the memory set. In the consistent-mapping condition, this could be done by simply judging whether the item was a number, which could be done by direct retrieval of a category label. In the varied-mapping condition, it was necessary to determine if the item was in the memory set. This did not require a sequential search but was done by a production pattern-match test whose time increased with the size of the set. This is analogous to the existing ACT model

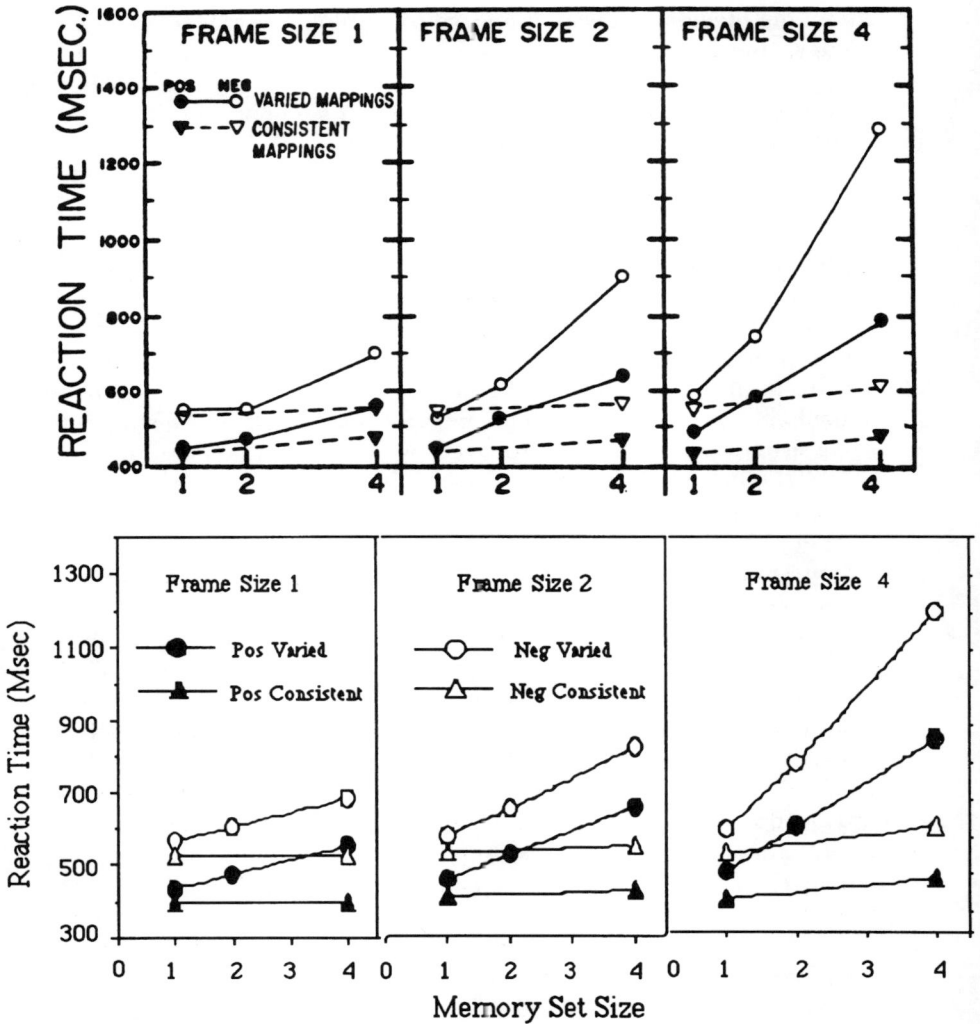

Figure 6. Results of Shiffrin and Schneider's (1977) simulation (top panel) and result of the ACT–R simulation (bottom panel). *Note.* Top panel from "Controlled and automatic human information processing: I. Detection, search and attention," by W. Schneider and R. M. Shiffrin, 1977, *Psychological Review, 84*(1), p. 19. Copyright © 1977 by the American Psychological Association.

for fan experiments and the Sternberg task (see Jones & Anderson, 1987). The basic productions for the two conditions are:

Judge-Consistent-Positive   **(BASE)**
    IF the goal is to search for an object with feature F
      and object attended is a number
THEN pop the goal and respond yes

Judge-Consistent-Negative   **(SHIFT)**
    IF the goal is to search for an object with feature F
      and object attended is a letter
THEN search for another object with feature F

Judge-Varied-Positive   **(BASE + i * FAN)**
    IF the goal is to search for an object with feature F
      and object attended is in the memory set
THEN pop the goal and respond yes

Judge-Varied-Negative   **(SHIFT + i*FAN)**
    IF the goal is to search for an object with feature F
      and object attended is not in the memory set
THEN search for another object with feature F

Terminate-No   **(BASE + NEG)**
    IF the goal is to search for an object with feature F
      and there are no unattended objects with feature F
THEN respond no

According to this model, the consistent-mapping condition enjoys two advantages over the varied-mapping condition. First, fewer positions will have to be examined because numbers have fewer features in common with letters than do letters. In the extreme condition (frame size = 4, set size = 4, negative trial), an average of 1.56 item positions had to be examined in the consistent condition and 2.49 item positions in the varied condition. The second advantage is that the target set did not have to be examined during judgment in the consistent condition, and so the condition did not suffer a fan effect. However, there still is a small effect of memory-set size in the consistent condition because there will be more target features to discriminate the targets from the letters as the memory-set size increases.

We developed a mathematical model of this ACT–R theory and fit it against Shiffrin and Schneider's (1977) data. This model required four parameters—a base reaction time (BASE, estimated to be 208 msec), an additional waiting time associated with **Terminate-No** for a negative response (NEG, estimated to be 133 msec), a time to attend to a position (SHIFT, estimated to be 186 msec), and a fan time per element (FAN, estimated to be 40 msec). The parameters associated with each production were given along with the productions. The predictions of the model are displayed in Figure 6 (bottom panel). The R^2 between the data and the predictions is .951, and the average mean deviation in prediction is 38 msec. The parameter values are also quite reasonable. Note the fan parameter is the slope in the typical Sternberg (1969) task. The time to attend to a position is reassuringly close to the estimate (185 msec) we obtained in fitting the Sperling task. Both the fan costs and the negative costs reflect the kinds of times we estimated previously for the effect of these factors on production matching.[6]

3.4. Conclusions

We have shown that the ACT–R model is consistent with some of the classic results from visual attention. In each of three tasks, we were able to fit the data assuming just about 185 msec to switch attention. In the Sperling task, attention switching was the only activity. In the subitizing task, there was also time required to set up and increment a count. In the Shiffrin and Schneider (1977) task, judgment time played a significant role. When we go to more cognitively loaded tasks, other processes will play still more significant roles. However, every time visual attention switches, approximately another 185 msec will be added to the processing time.

4. APPLICATION TO MENU-SELECTION DATA

We thought it would serve as a useful illustration to apply this ACT–R system to the same Nilsen (1991) data described by Kieras and Meyer (1994) in their report of the EPIC model. These data are concerned with time to scan a menu as a function of the target position of the item in the menu. The menu consists of a set of digits (1 to 9) randomly ordered vertically. The data to be modeled are the times for subjects to move a mouse from home position above the menu to the target item. Figure 7 shows the time for this action as a function of the serial position of the item in the menu. A linear function is obtained with a slope of 103 msec per position.

It is critical for this study that the items in the menu are ordered randomly. Because the subject does not know where the target item is, a critical component to latency has to be a serial search of the list looking for the target item. Subjects tend to move the mouse down as they scan for the target. Thus, after they identify the target, the distance to move the mouse tends not to vary much with serial position. Thus, our view is that, when the

---

6. An Excel file giving the fit can be found by following the path given in the Background note at the end of this article.

Figure 7. Observed and predicted menu-selection times. Observed data are from Nilsen (1991).

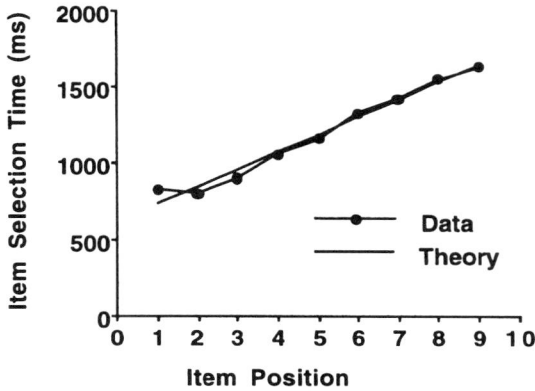

target position is unknown, time is dominated by visual search. In contrast, if the position of the item was known (as in a fixed order menu), the critical latency component might be a Fitts's-law description of the motion.

Our model for this task was essentially the same model as we proposed for Shiffrin and Schneider's (1977) data (Figure 6). We assume that, given a target, subjects selected one of its features and scanned down the menu for the first item with that feature. If this was the target, they stopped. If not, they scanned for the next item that contained the target feature.

The two critical productions are:

Hunt-Feature
> IF the goal is to find a target that has feature F
> and there is an unattended object below the current location with feature F
> THEN move attention to closest such object

Found-Target
> IF the goal is to find a target
> and the target is at location L
> THEN move the mouse to L and click

The first production **Hunt-Feature** moves attention down looking at objects that have the target feature. The movement of attention to an object will cause its identity to be encoded. If it is an instance of the target letter, **Found-Target** can apply. The production **Found-Target** will retrieve the location of the target and move the mouse to that location.

The time to reach a target will be a function of the number of digits that precede it that have the selected feature. It turns out, given the McClelland

Figure 8. Interaction between target and background.

Target	Background	
	Number[a]	Letter[a]
Number	1,324	1,293
Letter	1,253	1,366

[a]In milliseconds.

and Rumelhart (1981) feature set, there is a .53 probability that a randomly selected feature of one number will overlap with the feature set of another number.[7] Using the estimate (from Shiffrin & Schneider, 1977) of 186 msec for a shift of attention, we predict $186 \times .53 = 99$ msec per menu item, which is close to the slope (103 msec) in the Nilsen (1991) data. The fit of our model to the data is illustrated in Figure 7. This is a striking demonstration of how the ACT–R theory can be used to predict new data sets using old parameters.

Kieras and Meyer's (1994) EPIC model is able to an equally good job assuming a pipeline model whereby there are eye movements every 103 msec that will overrun the target. This strikes us as a very improbable speed of eye movement, which is conventionally set at about 200 msec. Kieras and Meyer suggested an alternative model in which as many as three items are processed in each gaze. Either of these models would predict no effect of distractor similarity on search time. In light of studies like Shiffrin and Schneider's (1977), this prediction seems unlikely. In contrast, the ACT–R model would predict, for instance, that it would be easier to find a number in a menu of letters than in a menu of numbers.

To test this prediction, we performed a within-subject menu-search task in which subjects had to select either a capital letter or a digit in a background of letters or digits. Figure 8 presents the results from subjects for menus of nine elements, as in Nilsen (1991). As predicted by ACT–R, subjects are significantly, $F(1, 20) = 104.77$, $p < .01$, faster when the distractors are different than the target. This is a confirmation of ACT–R's conception of visual attention and a token of its potential for modeling HCI tasks.

Although the interaction is to be expected, there is one unexpected result in the data—that there is a significant effect of background, with subjects slower (41 msec) in the presence of a letter background, $F(1, 20) = 29.96$, $p < .001$. We have no explanation of this effect.

---

7. Compared to Shiffrin and Schneider (1977; consistent-mapping condition), we did not assume that subjects had enough practice to select the most discriminating feature.

Figure 9. Observed and predicted selection times for numbers versus letters against a letter background. The predictions of the ACT–R theory are given in the solid lines.

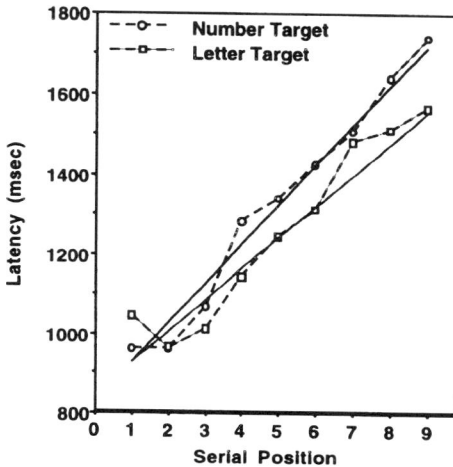

The strongest prediction of the ACT–R theory is that there should be a significant Serial Position × Target × Background interaction. In fact, there are significant Target × Position, $F(8, 160) = 6.49$, $p < .001$, Background × Position, $F(8, 160) = 4.30$, $p < .001$, and Target × Background × Position, $F(8, 168) = 2.18$, $p < .05$, interactions. There are significant differences among slopes, with 103 msec in the number-on-number condition, 84 msec in the number-on-letter condition, 80 msec in the letter-on-number condition, and 82 msec in the letter-on-letter condition. The basic effect is a steeper slope in the number-on-number condition. We calculated the mean probability of feature overlap in the conditions. There is a 53% probability overlap of the number-on-number condition, 39% in the number-on-letter condition, 42% in the letter-on-number condition, and 43% in the letter-on-letter condition. Thus, these overlap scores predict that there will be less ability to use features to guide search in the number-on-number condition. This prediction is confirmed.

Figure 9 plots the predictions of the ACT–R theory for number and letter targets holding constant the background as numbers. ACT–R is already committed as to the slopes in these cases. For number targets, it is $186 × .53 = 99$ msec (actual slope $= 103$); for letter targets, it is $186 × .42 = 78$ msec (actual slope $= 80$ msec). The only degree of freedom in estimating this is the "intercept" when the serial position is one. This was estimated as 927 msec. This is a striking confirmation of the ACT–R analysis of menu scanning in comparison to the EPIC model, which fails to predict these effects of Target × Background interaction. It also serves more generally to indicate the relevance of research on visual attention to HCI.

5. CONCLUSIONS

Our goal in this article has been to describe how we have given eyes to ACT–R. Although the model has a theory of visual perception, we have not concentrated on this but rather have focused on the important role of visual attention in accounting for data patterns. Elsewhere (Anderson & Douglass, 1998), we focused on the important role of visual attention in classic problem-solving tasks such as equation solving. However, here our goal has been to show that we properly model the basic processes of visual attention and that they matter in a traditional HCI task such as menu scanning. A critical value was the approximately 185 msec involved in shifting attention to an item in a visual array. However, if this value is simplisticly applied to a task, one can overestimate the time to shift attention because attention has the capacity to focus on items with specific features, and one needs to consider the implication of this focus for search time. For instance, Figure 9 shows that differential focus will result in differential search speed. ACT–R provides an architecture in which to work out these complex interactions with visual attention for both simple and complex tasks.

NOTES

Background. ACT–R and its visual interface can be accessed at its website: http://act.psy.cmu.edu/act/. This contains a po.nter to the ACT–R visual interface page, where the simulations and Excel parameter estimation files can be found.

Acknowledgments. We thank Chris Schunn for his comments on the research.

Support. We acknowledge Office of Naval Research Grant N00014–96–I–041 in supporting this research.

Authors' Present Addresses. John R. Anderson, Michael Matessa, and Christian Lebiere, Department of Psychology, Carnegie Mellon University, Pittsburgh, PA 15213. E-mail: ja+@cmu.edu; mm4b@andrew.cmu.edu; cl+@cmu.edu.

HCI Editorial Record. First manuscript received January 20, 1996. Revisions received June 18, 1996, and December 1996. Accepted by Wayne D. Gray. Final manuscript received February 17, 1997. — *Editor*

REFERENCES

Anderson, J. R. (1993). *Rules of the mind.* Hillsdale, NJ: Lawrence Erlbaum Associates, Inc.

Anderson, J. R., Corbett, A. T., Koedinger, K., & Pelletier, R. (1995). Cognitive tutors: Lessons learned. *Journal of the Learning Sciences, 4,* 167–207.

Anderson, J. R., & Douglass, S. (1998). *Visual attention and problem solving.* Manuscript in preparation.

Anderson, J. R., Reder, L. M., & Lebiere, C. (1996). Working memory: Activation limitations on retrieval. *Cognitive Psychology, 30,* 221–256.

Haber, R. N. (1983). The impending demise of the icon: A critique of the concept of iconic storage in visual information processing. *Behavioral and Brain Sciences, 6,* 1–11.

Jensen, E. M., Reese, E. P., & Reese, T. W. (1950). The subitizing and counting of visually presenting fields of dots. *Journal of Psychology, 30,* 363–392.

Jones, W. P., & Anderson, J. R. (1987). Short- and long-term memory retrieval: A comparison of the effects of information load and relatedness. *Journal of Experimental Psychology: General, 116,* 137–153.

Kieras, D. E., & Meyer, D. E. (1994). *The EPIC architecture for modeling human information-processing and performance: A brief introduction* (Report 1, TR–94/ONR–EPIC–1). Ann Arbor: University of Michigan, Department of Electrical Engineering.

Klein, R. (1988). Inhibitory tagging system facilitates visual search. *Nature, 334,* 430–431.

Lebiere, C. (1997). *An ACT–R model of cognitive arithmetic.* Dissertation proposal, School of Computer Science, Carnegie Mellon University, Pittsburgh, PA.

Mandler, G., & Shebo, B. J. (1982). Subitizing: An analysis of its component processes. *Journal of Experimental Psychology: General, 111,* 1–22.

McClelland, J. L., & Rumelhart, D. E. (1981). An interactive model of context effects in letter perception: I. An account of basic findings. *Psychological Review, 88,* 375–407.

Miller, G. A. (1956). The magical number seven, plus or minus two: Some limits on our capacity for processing information. *Psychological Review, 63,* 81–97.

Navon, D. (1977). Forest before trees: The precedence of global features in visual perception. *Cognitive Psychology, 9,* 353–383.

Nilsen, E. L. (1991). *Perceptual-motor control in human–computer interaction* (Technical Report 37). Ann Arbor: University of Michigan, Cognitive Science and Machine Intelligence Laboratory.

Posner, M. I. (1980). Orienting of attention. *Quarterly Journal of Experimental Psychology, 32,* 3–25.

Servan-Schreiber, E. (1991). *The competitive chunking theory: Models of perception, learning, and memory.* PhD dissertation, Department of Psychology, Carnegie Mellon University, Pittsburgh, PA.

Shiffrin, W., & Schneider, R. M. (1977). Controlled and automatic human information processing: I. Detection, search, and attention. *Psychological Review, 84,* 1–66.

Simon, T., Cabrera, A., & Kliegl, R. (1994). A new approach to the study of subitizing as distinct enumeration processing. *Proceedings of the sixteenth annual conference of the Cognitive Science Society,* 929–934. Hillsdale, NJ: Lawrence Erlbaum Associates, Inc.

Sperling, G. A. (1960). The information available in brief visual presentation. *Psychological Monographs, 74*(Whole No. 498).

Sternberg, S. (1969). Memory scanning: Mental processes revealed by reaction time experiments. *American Scientist, 57,* 421–457

Tipper, S. P., Driver, J., & Weaver, B. (1991). Object centered inhibition of return of visual attention. *Quarterly Journal of Experimental Psychology, 43A,* 289–298.

Treisman, A. M., & Gelade, G. (1980). A feature-integration theory of attention. *Cognitive Psychology, 12,* 97–136.

Treisman, A. M., & Sato, S. (1990). Conjunction search revisited. *Journal of Experimental Psychology: Human Perception and Performance, 16,* 459–478.

Wiesmeyer, M. D. (1992). *An operator-based model of covert visual attention.* Unpublished PhD thesis, Department of Computer Science, University of Michigan, Ann Arbor.

Wolfe, J. M. (1994). Guided search 2.0: A revised model of visual search. *Psychonomic Bulletin and Review, 1,* 202–238.

Acknowledgment to Reviewers

The quality of a journal can only be maintained by thoughtful, careful, and constructive reviewing. We thank the following people for taking the time to review manuscripts submitted to *Human-Computer Interaction* for Volume 12, 1997. *–Editor*

Gregory Abowd
Erik M. Altmann
Anne Anderson
John R. Anderson
Elisabeth Andre
Meera M. Blattner
Deborah A. Boehm-Davis
Ruven Brooks
Pam Burk
Michael Byrne
John M. Carroll
Jeff Conklin
Bill Curtis
Philip Cohen
Colleen Crangle
Stephanie Doane
Julie Dorsey
James D. Foley
Ephraim P. Glinert
Wayne D. Gray
Jonathan Grudin
Debbie Hindus
Ed Hovy
Andrew Howes
Ellen A. Isaacs
Robert J. K. Jacob

Bonnie E. John
Demetrios Karis
Irvin Katz
Joseph Kearney
Wendy Kellogg
David E. Kieras
Susan Kirschenbaum
Muneo Kitajima
Robert Kraut
Robin Lampert
Clayton Lewis
Jean MacMillan
Elisabeth Maier
Suzanne Mannes
Catherine Marshall
Mark Maybury
Jon Mayes
David Meader
Joseph S. Mertz, Jr.
Thomas P. Moran
Jon Oberlander
Gary Olson
Judith Olson
Mari Ostendorf
Sharon Oviatt

John W. Payne
Stephen Payne
Nancy Pennington
Richard Pew
Peter Pirolli
Catherine Plaisant
Peter G. Polson
Bob Rehder
John Rieman
Thomas Rist
Frank E. Ritter
Nadine Sarter
Dagmar Schmaucks
Oliviero Stock
Maureen Stone
J. Gregory Trafton
Carol-Ina Trudel
Wolfgang Wahlster
Joe Walther
Colin Ware
Suzi Weisband
Steve Whittaker
Cathy Wolf
Nicole Yankelovich
Richard M. Young

Author Index for Volume 12, 1997

HUMAN–COMPUTER INTERACTION

A Journal of Theoretical, Empirical and Methodological Issues of User Science and of System Design

Volume 12
1997

LEA LAWRENCE ERLBAUM ASSOCIATES, PUBLISHERS
Mahwah, New Jersey   London

HUMAN–COMPUTER INTERACTION

EDITOR: Thomas P. Moran, *Xerox Palo Alto Research Center*

ADMINISTRATIVE EDITOR: Patricia Sheehan, *Xerox Palo Alto Research Center*
PRODUCTION EDITOR: Joseph M. Coda, *Lawrence Erlbaum Associates, Inc.*

Human–Computer Interaction is published quarterly by Lawrence Erlbaum Associates, Inc., 10 Industrial Avenue, Mahwah, NJ 07430-2262.

Subscriptions (based on one volume per calendar year): Institutional, $215; Individual, $39. Subscriptions outside the United States and Canada: Institutional, $245; Individual, $69. Claims for missing issues cannot be honored beyond 4 months after mailing date. Duplicate copies cannot be sent to replace issues not delivered due to failure to notify publisher of change of address.

This journal is abstracted or indexed in: *Cambridge Scientific Abstracts: Health and Safety, Science Abstracts*; *Computer Abstracts*; *Educational Technology Abstracts*; *Engineering Information, Inc.*; *Ergonomics Abstracts*; *ERIC Clearinghouse in Education*; *ISI: SciSearch, Research Alert, Current Contents/Engineering, Computers & Technology*; *Knowledge Engineering Review*; *Library and Information Science Abstracts*; and *PsycINFO/Psychological Abstracts*. Microform copies of this journal are available through UMI, Periodical Check-In, North Zeeb Road, P.O. Box 1346, Ann Arbor, MI 48106–1346.

Printed in the United States   ISSN 0737–0024

HUMAN–COMPUTER INTERACTION

A Journal of Theoretical, Empirical, and Methodological Issues of User Science and of System Design

Contents of Volume 12

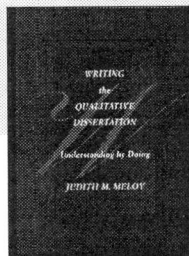

Subscription Order Form

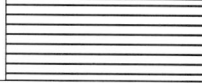

Please ❑ enter ❑ renew my subscription to

HUMAN-COMPUTER
INTERACTION
Volume 13, 1998, Quarterly

Subscription prices per volume:

Individual: ❑ $45.00 (US/Canada) ❑ $75.00 (All Other Countries)

Institution: ❑ $265.00 (US/Canada) ❑ $295.00 (All Other Countries)

Subscriptions are entered on a calendar-year basis only and must be prepaid in US currency -- check, money order, or credit card. **Offer expires 12/31/98. NOTE: Institutions must pay institutional rates.** Any institution paying for an individual subscription will be invoiced for the balance of the institutional subscription rate.

❑ Payment Enclosed

   Total Amount Enclosed $_____

❑ Charge My Credit Card

   ❑ VISA   ❑ MasterCard   ❑ AMEX   ❑ Discover

   Exp. Date_____

   Card Number _____

   Signature _____
   (Credit card orders cannot be processed without your signature.)

Please print clearly to ensure proper delivery.

Name _____

Address _____

_____

City _____ State _____ Zip+4 _____
Prices are subject to change without notice.

Lawrence Erlbaum Associates, Inc.
Journal Subscription Department
10 Industrial Avenue, Mahwah, NJ 07430
(201) 236-9500   FAX (201) 236-0072

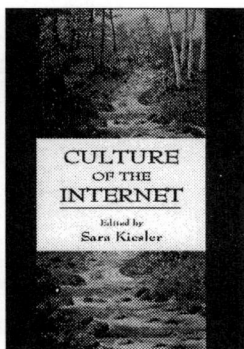

CULTURE
OF THE INTERNET
edited by
Sara Kiesler
Carnegie Mellon University

As the new century approaches, the astonishing spread of nationally and internationally accessible computer-based communication networks has touched the imagination of people everywhere. Suddenly, the Internet is in everyday parlance, featured in talk shows, in special business "technology" sections of major newspapers, and on the covers of national magazines. If the Internet is a new world of social behavior it is also a new world for those who study social behavior. This volume is a compendium of essays and research reports representing how researchers are thinking about the social processes of electronic communication and its effects in society. Taken together, the chapters comprise a first gathering of social psychological research on electronic communication and the Internet.

The authors of these chapters work in different disciplines and have different goals, research methods, and styles. For some, the emergence and use of new technologies represent a new perspective on social and behavioral processes of longstanding interest in their disciplines. Others want to draw on social science theories to understand technology. A third group holds to a more activist program, seeking guidance through research to improve social interventions using technology in domains such as education, mental health, and work productivity. Each of these goals has influenced the research questions, methods, and inferences of the authors and the "look and feel" of the chapters in this book.

Intended primarily for researchers who seek exposure to diverse approaches to studying the human side of electronic communication and the Internet, this volume has three purposes:

- to illustrate how scientists are thinking about the social processes and effects of electronic communication;
- to encourage research-based contributions to current debates on electronic communication design, applications, and policies; and
- to suggest, by example, how studies of electronic communication can contribute to social science itself.

0-8058-1635-6 [cloth] / 1997 / 480pp. / $99.95
0-8058-1636-4 [paper] / 1997 / 480pp. / $39.95

Lawrence Erlbaum Associates, Inc.
10 Industrial Avenue, Mahwah, NJ 07430

Prices subject to
change without notice.

201/236-9500   FAX 201/236-0072

Call toll-free to order: 1-800-9-BOOKS-9...9am to 5pm EST only.
e-mail to: orders@erlbaum.com
visit LEA's web site at http://www.erlbaum.com

LEA

VIDEO- MEDIATED COMMUNICATION

edited by
Kathleen E. Finn
Consultant
Abigail J. Sellen
Rank Xerox Research Centre
Sylvia B. Wilbur
Queen Mary and Westfield College,
University of London
A VOLUME IN THE COMPUTERS, COGNITION AND WORK SERIES

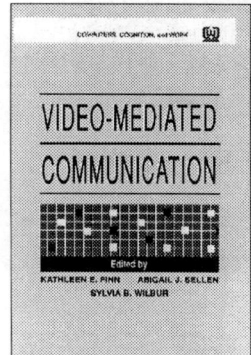

Decades after their introduction, video communication systems are beginning to realize their potential in supporting working from home, teaching and learning at a distance, conferencing, and interpersonal communication. In the face of an upsurge in interest, important questions are being asked: What function does video really serve, and what advantages over the telephone does it provide? How and why is video-mediated interaction different from face-to-face interaction? How can we best configure video technology to support different kinds of work at a distance? What is the role of video technology in the future?

People from a variety of disciplines have now produced a substantial body of research addressing these issues from a wide range of analytic perspectives. Their results and conclusions are scattered through journals, conference proceedings, and corporate technical papers. Drawing together the ideas and findings of the major researchers in the field, this volume offers the first comprehensive overview of what is currently known about video-mediated communication.

Written by psychologists, sociologists, anthropologists, engineers, and computer scientists, this book is an essential resource for all those who design and study systems for teaching, learning, and working. It is divided into four sections as follows:

- Foundations surveys the literature, constructs a foundational framework, introduces common vocabulary, and helps explain technical aspects and terms.
- Findings presents empirical work of types ranging from psychological laboratory-based studies to ethnographic field studies.
- Design explores various aspects of the design and evaluation of new kinds of video systems.
- The Future comments on new and innovative applications of video technology and points out the ways in which its use may be tied to broader technological trends.

0-8058-2288-7 [cloth] / 1997 / 584pp. / $99.95
Special Prepaid Offer! $49.95
No further discounts apply.

Lawrence Erlbaum Associates, Inc.
10 Industrial Avenue, Mahwah, NJ 07430
201/236-9500   FAX 201/236-0072

Prices subject to
change without notice.

Call toll-free to order: 1-800-9-BOOKS-9...9am to 5pm EST only.
e-mail to: orders@erlbaum.com
visit LEA's web site at http://www.erlbaum.com

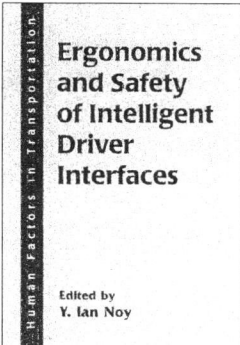

ERGONOMICS AND SAFETY OF INTELLIGENT DRIVER INTERFACES

Ergonomics and Safety of Intelligent Driver Interfaces

Edited by
Y. Ian Noy

edited by
Y. Ian Noy
Transport Canada
A VOLUME IN THE HUMAN FACTORS IN TRANSPORTATION SERIES

Even to the casual observer of the automotive industry, it is clear that driving in the 21st century will be radically different from driving as we know it today. Significant advances in diverse technologies such as digital maps, communication links, processors, image processing, chipcards, traffic management, and vehicle positioning and tracking, are enabling extensive development of intelligent transport systems (ITS). Proponents of ITS view these technologies as freeing designers to re-define the role and function of transport in society and to address the urgent problems of congestion, pollution, and safety. Critics, on the other hand, worry that ITS may prove too complex, too demanding, and too distracting for users, leading to loss of skill, increased incidence of human error, and greater risk of accidents.

The role of human factors is widely acknowledged to be critical to the successful implementation of such technologies. However, too little research is directed toward advancing the science of human-ITS interaction, and too little is published which is useful to system designers. This book is an attempt to fill this critical gap. It focuses on the intelligent driver interface (IDI) because the ergonomics of IDI design will influence safety and usability perhaps more than the technologies which underlie it.

The chapters cover a broad range of topics, from cognitive considerations in the design of navigation and route guidance, to issues associated with collision warning systems, to monitoring driver fatigue. The chapters also differ in intent — some provide design recommendations while others describe research findings or new approaches for IDI research and development. Based in part on papers presented at a symposium on the ergonomics of in-vehicle human systems held under the auspices of the 12th Congress of the International Ergonomics Association, the book provides an international perspective on related topics through inclusion of important contributions from Europe, North America, and Japan.

Many of the chapters discuss issues associated with navigation and route guidance because such systems are the most salient and arguably the most complex examples of IDI. However, the findings and research methodologies are relevant to other systems as well making this book of interest to a wide audience of researchers, design engineers, transportation authorities, and academicians involved with the development or implementation of ITS.

0-8058-1955-X [cloth] / 1997 / 448pp. / $89.95
0-8058-1956-8 [paper] / 1997 / 448pp. / $45.00

Lawrence Erlbaum Associates, Inc.

Prices subject to
change without notice.

10 Industrial Avenue, Mahwah, NJ 07430
201/236-9500   FAX 201/236-0072

Call toll-free to order: 1-800-9-BOOKS-9...9am to 5pm EST only.
e-mail to: orders@erlbaum.com
visit LEA's web site at http://www.erlbaum.com

LEA

Written in honor of Henil Stassen, one of the most prolific contributors to this research and literature, this book provides an up-to-date summary on human control of mechanical things. This includes people controlling the mechanical movements of their own limbs, extensions of their limbs such as prostheses (limb replacements), orthoses (limb braces), hand tools, or telemanipulators. It also consists of people controlling the mechanical movements of vehicles that they ride in such as aircraft, automobiles, and trains. Finally, it includes movements of discrete products through manufacturing plants, or chemicals and other fluids through process plants such as refineries or nuclear power stations.

Within academe or industry, these various types of human control are usually found in very different research and engineering communities. The first is generally regarded as a subfield of biomechanics and more generally of biomedical engineering — the hospital or medical clinic. Industry has mostly ignored the challenges of prosthetics and orthotics because there is not so much money to be made; payers are typically third party insurers of the government itself. In contrast, the problems of controlling vehicles (particularly aircraft and military vehicles) and industrial plants is what has driven the field of control — both the science and technology. In the field of robotics, where biomechanics is obviously a model, industry has experienced the effects of expecting too much too soon, and in some cases overinvesting and later being forced to withdraw in disappointment.

0-8058-2189-9 [cloth] / 1997 / 320pp. / $69.95
0-8058-2190-2 [paper] / 1997 / 320pp. / $39.95

Lawrence Erlbaum Associates, Inc.
10 Industrial Avenue, Mahwah, NJ 07430
201/236-9500   FAX 201/236-0072

Prices subject to
change without notice.

Call toll-free to order: 1-800-9-BOOKS-9...9am to 5pm EST only.
e-mail to: orders@erlbaum.com
visit LEA's web site at http://www.erlbaum.com

LEA